A First Course in Topology

Continuity and Dimension

Continuity
and Dimension

STUDENT MATHEMATICAL LIBRARY
Volume 31

A First Course in Topology

Continuity and Dimension

John McCleary

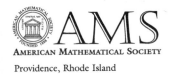

AMERICAN MATHEMATICAL SOCIETY

Providence, Rhode Island

Editorial Board

2000 *Mathematics Subject Classification*. Primary 54-01, 55-01, 54F45.

For additional information and updates on this book, visit
www.ams.org/bookpages/stml-31

Library of Congress Cataloging-in-Publication Data
McCleary, John, 1952-
   A first course in topology : continuity and dimension / John McCleary.
      p. cm. – (Student mathematical library, ISSN 1520-9121 ; v. 31)
   Includes bibliographical references.
   ISBN 0-8218-3884-9 (alk. paper)
   1. Topology–Textbooks. I. Title. II. Series.

QA611 .M38   2006
514–dc22
                                                          2005058915

Contents

Introduction

In the first place, what are the properties of space properly so called? ... 1st, it is continuous; 2nd, it is infinite; 3rd, it is of three dimensions; ...

HENRI POINCARÉ, 1905

So will the final theory be in 10, 11 *or* 12 *dimensions?*

MICHIO KAKU, 1994

As a separate branch of mathematics, topology is relatively young. It was isolated as a collection of methods and problems by HENRI POINCARÉ (1854–1912) in his pioneering paper *Analysis situs* of 1895. The subsequent development of the subject was dramatic and topology was deeply influential in shaping the mathematics of the twentieth century and today.

So what is topology? In the popular understanding, objects like the Möbius band, the Klein bottle, and knots and links are the first to be mentioned (or maybe the second after the misunderstanding about topography is cleared up). Some folks can cite the joke that topologists are mathematicians who cannot tell their donut from their coffee cups. When I taught my first undergraduate courses in topology, I found I spent too much time developing a hierarchy of definitions and too little time on the objects, tools, and intuitions that are central to the subject. I wanted to teach a course that would follow a path more directly to the heart of topology. I wanted to tell a story that is coherent, motivating, and significant enough to form the basis for future study.

To get an idea of what is studied by topology, let's examine its prehistory, that is, the vague notions that led Poincaré to identify its foundations. GOTTFRIED W. LEIBNIZ (1646–1716), in a letter to CHRISTIAAN HUYGENS (1629–1695) in the 1670's, described a concept that has become a goal of the study of topology:

> *I believe that we need another analysis properly geometric or linear, which treats PLACE directly the way that algebra treats MAGNITUDE.*

Leibniz envisioned a calculus of figures in which one might combine figures with the ease of numbers, operate on them as one might with polynomials, and produce new and rigorous geometric results. This science of *PLACE* was to be called *Analysis situs* ([**68**]).

We don't know what Leibniz had in mind. It was LEONHARD EULER (1701–1783) who made the first contributions to the infant subject, which he preferred to call *geometria situs*. His solution to the Bridges of Königsberg problem and the celebrated *Euler formula*, $V - E + F = 2$ (Chapter 11), were results that depended on the relative positions of geometric figures and not on their magnitudes ([**68**], [**46**]).

In the nineteenth century, CARL-FRIEDRICH GAUSS (1777–1855) became interested in *geometria situs* when he studied knots and links as generalizations of the orbits of planets ([**25**]). By labeling figures of knots and links Gauss developed a rudimentary calculus that distinguished certain knots from each other by combinatorial means. Students who studied with Gauss and went on to develop some of the threads associated with *geometria situs* were JOHANN LISTING (1808–1882), AUGUSTUS MÖBIUS (1790–1868), and BERNHARD RIEMANN (1826–1866). Listing extended Gauss's informal census of knots and links and he coined the term *topology* (from the Greek τοπου λογος, which in Latin is *analysis situs*). Möbius extended Euler's formula to surfaces and polyhedra in three-space. Riemann identified the methods of the emerging *analysis situs* as fundamental in the study of complex functions.

During the nineteenth century analysis was developed into a deep and subtle science. The notions of continuity of functions and the convergence of sequences were studied in increasingly general situations, especially in the work of GEORG CANTOR (1845–1918) and finalized in the twentieth century by FELIX HAUSDORFF (1869–1942) who proposed the general notion of a *topological space* in 1914 ([**33**]).

The central concept in topology is continuity, defined for functions between sets equipped with a notion of nearness (topological spaces) which is preserved by a continuous function. Topology is a kind of geometry in which the important properties of a figure are those that are preserved under continuous motions (homeomorphisms, Chapter 2). The popular image of topology as *rubber sheet geometry* is captured in this characterization. Topology provides a language of continuity that is general enough to include a vast array of phenomena while being precise enough to be developed in new ways.

A motivating problem from the earliest struggles with the notion of continuity is the problem of dimension. In modern physics, higher dimensional manifolds play a fundamental role in describing theories with properties that combine the large and the small. Already in Poincaré's time the question of the physicality of dimension was on philosophers' minds, including Poincaré. Cantor had noticed in 1877 that, as sets, finite-dimensional Euclidean spaces were indistinguishable (Chapter 1). If these identifications were possible in a continuous manner, a requirement of physical phenomena, then the role of dimension would need a critical reappraisal. The problem of dimension was important to the development of certain topological notions, including a strictly topological definition of dimension introduced by HENRI LEBESGUE (1875–1941) [**47**]. The solution to the problem of dimension was found by L. E. J. BROUWER (1881–1966) and published in 1910 [**10**]. The methods introduced by Brouwer reshaped the subject.

The story I want to tell in this book is based on the problem of dimension. This fundamental question from the early years of the subject organizes the exposition and provides the motivation for the choices of mathematical tools to develop. I have not chosen to follow the path of Lebesgue into dimension theory (see the classic text [38]) but the further ranging path of Poincaré and Brouwer. The fundamental group (Chapters 7 and 8) and simplicial methods (Chapters 10 and 11) provide tools that establish an approach to topological questions that has proven to be deep and is still developing. It is this approach that best fits Leibniz's wish.

In what follows, we will cut a swath through the varied and beautiful landscape that is the field of topology with the goal of solving the problem of invariance of dimension. Along the way we will acquire the necessary vocabulary to make our way easily from one landmark to the next (without staying too long anywhere to pick up an accent). The first chapter reviews the set theory with which the problem of dimension can be posed. The next five chapters treat the basic point-set notions of topology; these ideas are closest to analysis, including connectedness and compactness. The next two chapters treat the fundamental group of a space, an idea introduced by Poincaré to associate a group to a space in such a way that equivalent spaces lead to isomorphic groups. The next chapter treats the Jordan Curve Theorem, first stated by Jordan in 1882, and given a complete proof in 1905 by OSWALD VEBLEN (1880–1960). The method of proof here mixes the point-set and the combinatorial to develop approximations and comparisons. The last two chapters take up the combinatorial theme and focus on simplicial complexes. To these conveniently constructed spaces we associate their homology, a sequence of vector spaces, which turn out to be isomorphic for equivalent complexes. This leads to a proof of the topological invariance of dimension using homology.

Though the motivation for this book is historical, I have not followed the history in the choice of methods or proofs. First proofs of significant results can be difficult. However, I have tried to imitate the mix of point-set and combinatorial ideas that was topology before 1935, what I call classical topology. Some beautiful results of this time are included, such as the Borsuk-Ulam theorem (see [9] and [56]).

HOW TO USE THIS BOOK

I have tried to keep the prerequisites for this book at a minimum. Most students meeting topology for the first time are old hands at linear algebra, multivariable calculus, and real analysis. Although I introduce the fundamental group in Chapters 7 and 8, the assumptions I make about experience with groups are few and may be provided by the instructor or picked up easily from any book on modern algebra. Ideally, a familiarity with groups makes the reading easier, but it is not a hard and fast prerequisite.

A one-semester course in topology with the goal of proving Invariance of Dimension can be built on Chapters 1–8, 10, and 11. A stiff pace is needed for most undergraduate classes to get to the end. A short cut is possible by skipping Chapters 7 and 8 and focusing the end of the semester on Chapters 10 and 11. Alternatively, one could cover Chapters 1–8 and simply explain the argument of Chapter 11 by analogy with the case discussed in Chapter 8. Another suggestion is to make Chapter 1 a reading assignment for advanced students with a lot of experience with basic set theory. Chapter 9 is a classical result whose proof offers a bridge between the methods of Chapters 1–8 and the combinatorial emphasis of Chapters 10 and 11. This can be made into another nice reading assignment without altering the flow of the exposition.

For the undergraduate reader with the right background, this book offers a glimpse into the standard topics of a first course in topology, motivated by historically important results. It might make a good read in those summer months before graduate school.

Finally, for any gentle reader, I have tried to make this course both efficient in exposition and motivated throughout. Though some of the arguments require developing many interesting propositions, keep on the trail and I promise a rich introduction to the landscape of topology.

ACKNOWLEDGEMENTS

This book grew out of the topology course I taught at Vassar College off and on since 1989. I thank the many students who have taken it and who helped me in refining the arguments and emphases. Most recently, HeeSook Park taught the course from the manuscript and her questions, criticisms, and recommendations have been insightful; the text is better for her close reading. Molly Kelton improved the text during a reading course in which she questioned every argument closely. Conversations with Bill Massey, Jason Cantarella, Dave Ellis, and Sandy Koonce helped shape the organization I chose here. I learned the bulk of the ideas in the book first from Hugh Albright and Sam Wiley as an undergraduate, and from Jim Stasheff as a graduate student. My teachers taught me the importance and excitement of topological ideas—a gift for my life. I hope I have transmitted some of their good teaching to the page. I thank Dale Johnson for sharing his papers on the history of the notion of dimension with me. His work is a benchmark in the history of mathematics, and informed my account in the book. I thank Sergei Gelfand who has shepherded this project from conception to completion—his patience and good cheer are much appreciated. Also thanks to Mary Letourneau for eagle-eyed copy-editing and improvements of my prose. Finally, my thanks to my family, Carlie, John, and Anthony, for their patient support of my work.

While an undergraduate struggling with open and closed sets, I shared a house with friends who were a great support through all those years of personal growth. We called our house *Igorot*. This book is dedicated to my fellow *Igorots* (elected and honorary) who were with me then, and remained good friends so many years later.

Chapter 1

A Little Set Theory

Functions are the single most important idea pervading modern mathematics. We will assume the informal definition of a function—a well-defined rule assigning to each element of the set A a unique element in the set B. We denote these data by $f\colon A \to B$ and the rule by $f\colon a \in A \mapsto f(a) \in B$. The set A is the **domain** of f and the receiving set B is its **codomain** (or range). We make an important distinction between the range and the **image** of a function, $f(A) = \{f(a) \in B \mid a \in A\}$, which is a subset contained in B.

When the codomain of one function and the domain of another coincide, we can compose them, that is, given $f\colon A \to B$, $g\colon B \to C$ we can define $g \circ f\colon A \to C$ by the rule $g \circ f(a) = g(f(a))$. If $X \subset A$, then we write $f|_X\colon X \to B$ for the restriction of the rule of f to the elements of X. This changes the domain and so it is a different function. Another way to express $f|_X$ is to define the *inclusion function*

$$i\colon X \to A, \quad i(x) = x.$$

We can then write $f|_X = f \circ i\colon X \to B$.

Certain properties of functions determine the notion of equivalence of sets.

Definition 1.1. A function $f\colon A \to B$ is **one-one** (or injective) if whenever $f(a_1) = f(a_2)$, then $a_1 = a_2$. A function $f\colon A \to B$ is **onto** (or surjective) if for any $b \in B$, there is an $a \in A$ with $f(a) = b$. The function f is a **one-one correspondence** (or bijective, or an isomorphism of sets) if f is both one-one and onto. Two sets are **equivalent** or have the **same cardinality** if there is a one-one correspondence $f\colon A \to B$.

If $f\colon A \to B$ is a one-one correspondence, then f has an inverse function $f^{-1}\colon B \to A$. The inverse function is determined by the fact that if $b \in B$, then there is an element $a \in A$ with $f(a) = b$. Furthermore, a is uniquely determined by b because $f(a) = f(a') = b$ implies that $a = a'$. So we define $f^{-1}(b) = a$. It follows that $f \circ f^{-1}\colon B \to B$ is the identity mapping $\mathrm{id}_B(b) = b$, and likewise $f^{-1} \circ f\colon A \to A$ is the identity id_A on A.

For example, if we restrict the tangent function of trigonometry to $(-\pi/2, \pi/2)$, then we get a one-one correspondence $\tan\colon (-\pi/2, \pi/2) \to \mathbb{R}$. The inverse function is the arctan function. Furthermore, any open interval (a, b) is equivalent to any other (c, d) via the one-one correspondence $t \mapsto c + [(d - c)(t - a)/(b - a)]$. Thus the set of real numbers is equivalent as sets to any open interval of real numbers.

Given a function $f\colon A \to B$, we can define new functions on the collections of subsets of A and B. For any set S, let $\mathcal{P}(S) = \{X \mid X \subset S\}$ denote the **power set** of S. We define the **image** of a subset $X \subset A$ by

$$f(X) = \{f(x) \in B \mid x \in X\},$$

and this determines a function $f\colon \mathcal{P}(A) \to \mathcal{P}(B)$. Define the **preimage** of a subset $U \subset B$ by

$$f^{-1}(U) = \{x \in A \mid f(x) \in U\}.$$

The preimage determines a function $f^{-1} \colon \mathcal{P}(B) \to \mathcal{P}(A)$. This is a splendid abuse of notation, however, *don't confuse the preimage with an inverse function*. Inverse functions only exist when f is one-one and onto. Furthermore, the domain of the preimage is the set of subsets of B. We list some properties of the image and preimage functions. The proofs are left to the reader.

Proposition 1.2. *Let $f \colon A \to B$ be a function and U, V subsets of B. Then*

1) *If $U \subset V$, then $f^{-1}(U) \subset f^{-1}(V)$.*
2) $f^{-1}(U \cup V) = f^{-1}(U) \cup f^{-1}(V)$.
3) $f^{-1}(U \cap V) = f^{-1}(U) \cap f^{-1}(V)$.
4) $f(f^{-1}(U)) \subset U$.
5) *For $X \subset A$, $X \subset f^{-1}(f(X))$.*
6) *If, for any $U \subset B$, $f(f^{-1}(U)) = U$, then f is onto.*
7) *If, for any $X \subset A$, $f^{-1}(f(X)) = X$, then f is one-one.*

Equivalence relations

A significant notion in set theory is the **equivalence relation**. A relation, R, is formally a subset of the set of pairs $A \times A$, of a set A. We write $x \sim y$ whenever $(x, y) \in R$.

Definition 1.3. A relation $\sim$ is an **equivalence relation** if

1) For all x in A, $x \sim x$. (Reflexive)
2) If $x \sim y$, then $y \sim x$. (Symmetric)
3) If $x \sim y$ and $y \sim z$, then $x \sim z$. (Transitive)

Examples. (1) For any set A, the relation of equality $=$ is an equivalence relation: No element is related to any other element except itself.

(2) Let $A = \mathbb{Z}$, the set of integers with the usual sense of divisibility. Given a nonzero integer m, write $k \equiv l$ whenever m divides $l - k$, denoted $m \mid l - k$. Notice that $m \mid 0 = k - k$ so $k \equiv k$ for any k and $\equiv$ is reflexive. If $m \mid l - k$, then $m \mid -(l - k) = k - l$ so that $k \equiv l$ implies $l \equiv k$ and $\equiv$ is symmetric. Finally, suppose

for some integers d and e that $l - k = md$ and $j - l = me$. Then $j - k = j - l + l - k = me + md = m(e + d)$. This shows that $k \equiv l$ and $l \equiv j$ imply $k \equiv j$ and $\equiv$ is transitive. Thus $\equiv$ is an equivalence relation. It is usual to write $k \equiv l \pmod{m}$ to keep track of the dependence on m.

(3) Let $\mathcal{P}(A) = \{U \mid U \subset A\}$ denote the power set of A. Then we can define a relation $U \leftrightarrow V$ whenever there is a one-one correspondence $U \to V$. The identity function $\mathrm{id}_U \colon U \to U$ establishes that $\leftrightarrow$ is reflexive. The fact that the inverse of a one-one correspondence is also a one-one correspondence proves $\leftrightarrow$ is symmetric. Finally, the composition of one-one correspondences is a one-one correspondence and so $\leftrightarrow$ is transitive. Thus $\leftrightarrow$ is an equivalence relation.

(4) Suppose $B \subset A$. Then we can define a relation by $x \sim y$ if x and y are both in B; otherwise, $x \sim y$ only if $x = y$. This relation comes in handy later (Chapter 4).

Given an equivalence relation on a set A, say $\sim$, we define the **equivalence class** of an element a in A by

$$[a] = \{b \in A \mid a \sim b\}.$$

We denote the set of equivalence classes $[A] = \{[a] \mid a \in A\}$. Finally, let p denote the mapping $p \colon A \to [A]$ given by $p(a) = [a]$.

Proposition 1.4. *If a, $b \in A$, then as subsets of A, either $[a] = [b]$, when $a \sim b$, or $[a] \cap [b] = \emptyset$.*

Proof. If $c \in [a] \cap [b]$, then $a \sim c$ and $b \sim c$. By symmetry we have $c \sim b$ and so, by transitivity, $a \sim b$. Suppose $x \in [a]$; then $x \sim a$, and with $a \sim b$ we have $x \sim b$ and $x \in [b]$. Thus $[a] \subset [b]$. Reversing the roles of a and b in this argument we get $[b] \subset [a]$ and so $[a] = [b]$. $\square$

This proposition shows that the equivalence classes of an equivalence relation on a set A partition the set into disjoint subsets. The canonical function $p \colon A \to [A]$ has special properties.

Proposition 1.5. *The function $p \colon A \to [A]$ is a surjection. If $f \colon A \to Y$ is any other function for which, whenever $x \sim y$ in A, we have $f(x) = f(y)$, then there is a mapping $\overline{f} \colon [A] \to Y$ for which $f = \overline{f} \circ p$.*

Proof. The surjectivity of p is immediate. To construct $\overline{f}\colon [A] \to Y$ let $[a] \in [A]$ and define $\overline{f}([a]) = f(a)$. We need to check that this rule is well defined. Suppose $[a] = [b]$. Then we require $f(a) = f(b)$. But this follows from the condition that $a \sim b$ implies $f(a) = f(b)$. To complete the proof, $\overline{f}([a]) = \overline{f}(p(a)) = f(a)$ and so $f = \overline{f} \circ p$. $\qquad\square$

Of course, $p^{-1}([a]) = \{b \in A \mid b \sim a\} = [a]$ *as a subset of A, not as an element of the set* $[A]$. We have already observed that the equivalence classes partition A into disjoint pieces. Equivalently suppose $P = \{C_\alpha, \alpha \in I\}$ is a collection of subsets that partitions A, that is,

$$\bigcup_{\alpha \in I} C_\alpha = A \quad \text{and} \quad C_\alpha \cap C_\beta = \emptyset \ \text{ if } \alpha \neq \beta.$$

We can define a relation on A from the partition by

$$x \sim_P y \text{ if there is an } \alpha \in I \text{ with } x, y \in C_\alpha.$$

Proposition 1.6. *The relation $\sim_P$ is an equivalence relation. Furthermore there is a one-one correspondence between $[A]$ and P.*

Proof. $x \sim_P x$ follows from $\bigcup_{\alpha \in I} C_\alpha = A$. Symmetry and transitivity follow easily. The one-one correspondence required for the isomorphism is given by

$$f\colon A \to P \quad \text{where } a \mapsto C_\alpha, \text{ if } a \in C_\alpha.$$

By Proposition 1.5 this factors as a mapping $\overline{f}\colon [A] \to P$, which is onto. We check that $\overline{f}$ is one-one: if $\overline{f}([a]) = \overline{f}([b])$, then $a, b \in C_\alpha$ for the same α and so $a \sim_P b$, which implies $[a] = [b]$. $\qquad\square$

This discussion leads to the following equivalence of sets:

$$\{\text{Partitions of a set } A\} \Longleftrightarrow \{\text{Equivalence relations on } A\}.$$

Sets like the integers $\mathbb{Z}$ or a vector space V enjoy extra structure—you can add and subtract elements. You also can multiply elements in $\mathbb{Z}$, or multiply by scalars in V. When there is an equivalence relation on sets with the extra structure of a binary operation one can ask if the relation respects the operation. We consider two important examples and then deduce general conditions for this special property.

Example 1. For the equivalence relation $\equiv \pmod{m}$ on $\mathbb{Z}$ with $m \neq 0$ it is customary to write

$$[\mathbb{Z}] =: \mathbb{Z}/m\mathbb{Z}.$$

Given two equivalence classes in $\mathbb{Z}/m\mathbb{Z}$, can we add them to get another? The most obvious idea to try is the following formula:

$$[i] + [j] = [i + j].$$

To be sure this makes sense, remember $[i] = [i']$ whenever $i \equiv i' \pmod{m}$ so we have to be sure any changes of representative of an equivalence class do not alter the sum equivalence class. Suppose $[i] = [i']$ and $[j] = [j']$; then we require $[i + j] = [i' + j']$ if we want a definition of $+$ on $\mathbb{Z}/m\mathbb{Z}$. Let $i' - i = rm$ and $j' - j = sm$; then

$$i' + j' - (i + j) = (i' - i) + (j' - j) = rm + sm = (r + s)m$$

or $m \mid (i' + j') - (i + j)$, and so $[i + j] = [i' + j']$. Subtraction is also well defined on $\mathbb{Z}/m\mathbb{Z}$ and the element $0 = [0]$ acts as an additive identity in $\mathbb{Z}/m\mathbb{Z}$. Thus $\mathbb{Z}/m\mathbb{Z}$ has the structure of a group. It is a finite group given as the set

$$\mathbb{Z}/m\mathbb{Z} = \{[0], [1], [2], \ldots, [m - 1]\}.$$

Example 2. Suppose W is a linear subspace of a finite-dimensional vector space V. Define a relation on V by $u \equiv v \pmod{W}$ whenever $v - u \in W$. We check that we have an equivalence relation:

reflexive: If $v \in V$, then $v - v = 0 \in W$, since W is a subspace.

symmetric: If $u \equiv v \pmod{W}$, then $v - u \in W$ and so

$$(-1)(v - u) = u - v \in W$$

since W is closed under multiplication by scalars. Thus $v \equiv u \pmod{W}$.

transitive: If $u \equiv v \pmod{W}$ and $v \equiv x \pmod{W}$, then $x - v$ and $v - u$ are in W. Then $x - v + v - u = x - u$ is in W since W is a subspace. So $u \equiv x \pmod{W}$.

We denote $[V]$ as V/W. We next show that V/W is also a vector space. Given $[u], [v]$ in V/W, define $[u] + [v] = [u + v]$ and $c[u] = [cu]$. To see that these operations are well defined, suppose $[u] = [u']$ and $[v] = [v']$. We compare $(u' + v') - (u + v)$. Since $u' - u \in W$ and $v' - v \in W$, we have $(u' + v') - (u + v) = (u' - u) + (v' - v)$ is in W.

Similarly, if $[u] = [u']$, then $u' - u \in W$ so $c(u' - u) = cu' - cu$ is in W and $[cu] = [cu']$. The other axioms for a vector space hold in V/W by heredity and so V/W is a vector space. The canonical mapping $p: V \to V/W$ is a linear mapping:

$$p(cu + c'v) = [cu + c'v] = [cu] + [c'v]$$
$$= c[u] + c'[v] = cp(u) + c'p(v).$$

The kernel of the mapping is $p^{-1}([0]) = W$. Thus the dimension of V/W is given by

$$\dim V/W = \dim V - \dim W.$$

This construction is very useful and appears again in Chapter 11.

A general result applies to a set A with a binary operation $\mu: A \times A \to A$ and an equivalence relation on A.

Definition 1.7. An equivalence relation $\sim$ on a set A with binary operation $\mu: A \times A \to A$ is a **congruence relation** if the mapping $\bar{\mu}: [A] \times [A] \to [A]$ given by

$$\bar{\mu}([a], [b]) = [\mu(a, b)]$$

induces a well-defined binary operation on $[A]$.

The operation of $+$ on $\mathbb{Z}$ is a congruence relation with respect to the equivalence relation $\equiv$ (mod m). The operation of $+$ is a congruence relation on a vector space V with respect to the equivalence relation induced by a subspace W. More generally, well-definedness is the important issue in identifying a congruence relation.

Proposition 1.8. *An equivalence relation $\sim$ on A with $\mu: A \times A \to A$ is a congruence relation if for any $a, a', b, b' \in A$, whenever $[a] = [a']$ and $[b] = [b']$, we have $[\mu(a, b)] = [\mu(a', b')]$.*

The Schröder-Bernstein Theorem

There is a marvelous criterion for the existence of a one-one correspondence between two sets.

The Schröder-Bernstein Theorem. *If there are one-one mappings*

$$f\colon A \to B \text{ and } g\colon B \to A,$$

then there is a one-one correspondence between A and B.

Proof. In order to prove this theorem, we first prove the following preliminary result.

Lemma 1.9. *If $B \subset A$ and $f\colon A \to B$ is one-one, then there exists a function $h\colon A \to B$, which is a one-one correspondence.*

Proof ([14]). Take $B \subset A$ and suppose $B \neq A$. Recall that $A - B = \{a \in A \mid a \notin B\}$. Define

$$C = \bigcup_{n \geq 0} f^n(A - B),$$

where $f^0 = \mathrm{id}_A$ and $f^k(x) = f\big(f^{k-1}(x)\big)$. Define the function $h\colon A \to B$ by

$$h(a) = \begin{cases} f(a), & \text{if } a \in C, \\ a, & \text{if } a \in A - C. \end{cases}$$

By definition, $A - B \subset C$ and $f(C) \subset C$. Suppose $n > m \geq 0$. Observe that

$$f^m(A - B) \cap f^n(A - B) = \emptyset.$$

To see this suppose $f^m(x) = f^n(x')$. Then $f^{n-m}(x') = x \in A - B$. But $f^{n-m}(x') \in B$ and so $x \in (A - B) \cap B = \emptyset$, a contradiction. This implies that h is one-one, since f is one-one.

We next show that h is onto:

$$h(A) = f(C) \cup (A - C)$$

$$= f\left(\bigcup_{n \geq 0} f^n(A - B)\right) \cup \left(A - \bigcup_{n \geq 0} f^n(A - B)\right)$$

$$= \bigcup_{n \geq 1} f^n(A - B) \cup \left(A - \bigcup_{n \geq 0} f^n(A - B)\right)$$

$$= A - (A - B) = B.$$

So h is a one-one correspondence. $\qquad\square$

Proof of the Schröder-Bernstein Theorem. Let $A_0 = g(B) \subset A$ and $B_0 = f(A) \subset B$. Then $g_0\colon B \to A_0$ and $f_0\colon A \to B_0$ are one-one correspondences, each induced by g and f, respectively. Let

$F = f_0 \circ g_0 \colon B \to B_0$ denote the one-one function. Lemma 1.9 applies to (B, B_0, F), so there is a one-one correspondence $h \colon B_0 \to B$. The composition $h \circ f_0 \colon A \to B_0 \to B$ is the desired equivalence of sets. $\quad\square$

The problem of Invariance of Dimension

The development of set theory brought new insights about infinity. In particular, a set and its power set have different cardinalities. When a set is infinite, the cardinality of the power set is greater, and so there is a hierarchy of infinities. The discovery of this hierarchy prompted Cantor, in his correspondence with RICHARD DEDEKIND (1831–1916), to ask whether higher-dimensional sets might be distinguished by cardinality. On 5 January 1874 Cantor wrote to Dedekind and posed the question:

> Can a surface (perhaps a square including its boundary) be put into one-one correspondence with a line (perhaps a straight line segment including its endpoints) ... ?

He was soon able to prove the following positive result.

Theorem 1.10. *There is a one-one correspondence* $\mathbb{R} \to \mathbb{R} \times \mathbb{R}$.

Proof. We apply the Schröder-Bernstein Theorem. Since the mapping $f \colon \mathbb{R} \to (0,1)$ given by $f(r) = \frac{1}{\pi}\left(\arctan(r) + \frac{\pi}{2}\right)$ is a one-one correspondence, it suffices to show that there is a one-one correspondence between $(0,1)$ and $(0,1) \times (0,1)$. We obtain one assumption of the Schröder-Bernstein theorem because there is a one-one mapping $f \colon (0,1) \to (0,1) \times (0,1)$ given by the diagonal mapping $f \colon t \mapsto (t,t)$.

To apply the Schröder-Berstein Theorem we construct an injection $(0,1) \times (0,1) \to (0,1)$. Recall that every real number can be expressed as a *continued fraction* ([**31**]): suppose $r \in \mathbb{R}$. The **least integer function** (or *floor function*) is defined by

$$\lfloor r \rfloor = \max\{j \in \mathbb{Z} \mid j \leq r\}.$$

Since $0 < r < 1$, it follows that $1/r > 1$. Let $a_1 = \lfloor 1/r \rfloor$ and $r_1 = (1/r) - \lfloor 1/r \rfloor$. Then $0 \le r_1 < 1$. We can write

$$r = \frac{1}{\dfrac{1}{r}} = \frac{1}{\dfrac{1}{r} - \left\lfloor \dfrac{1}{r} \right\rfloor + \left\lfloor \dfrac{1}{r} \right\rfloor} = \frac{1}{a_1 + r_1}.$$

If $r_1 = 0$ we can stop. If $r_1 > 0$, then repeat the process to r_1 to obtain a_2 and r_2 for which

$$r = \cfrac{1}{a_1 + \cfrac{1}{a_2 + r_2}}.$$

Continuing in this manner, we can express r as a continued fraction:

$$r = \cfrac{1}{a_1 + \cfrac{1}{a_2 + \cfrac{1}{a_3 + \cdots}}} =: [0; a_1, a_2, a_3, \ldots].$$

For example,

$$\frac{31}{127} = \cfrac{1}{4 + \cfrac{3}{31}} = \cfrac{1}{4 + \cfrac{1}{10 + \cfrac{1}{3}}} = [0; 4, 10, 3].$$

We can recognize a rational number by the fact that its continued fraction terminates after finitely many steps. Irrationals have infinite continued fractions, for example, $1/\sqrt{2} = [0; 1, 2, 2, 2, \ldots]$.

To prove Cantor's theorem, we first introduce an injection $I \colon (0, 1) \to (0, 1)$ defined on continued fractions by

$$I(r) = \begin{cases} [0; a_1 + 2, a_2 + 2, \ldots, a_n + 2, 2, 2, \ldots], \\ \qquad\qquad \text{if } r = [0; a_1, a_2, \ldots, a_n], \\ [0; a_1 + 2, a_2 + 2, a_3 + 2, \ldots], \\ \qquad\qquad \text{if } r = [0; a_1, a_2, a_3, \ldots]. \end{cases}$$

Thus I maps all of the real numbers in $(0, 1)$ to the set $J = (0, 1) \cap (\mathbb{R} - \mathbb{Q})$ of irrational numbers in $(0, 1)$. We can define another one-one function $t \colon J \times J \to (0, 1)$, given by

$$t([0; a_1, a_2, \ldots], [0; b_1, b_2, \ldots]) = [0; a_1, b_1, a_2, b_2, \ldots].$$

The uniqueness of the continued fraction representation of a real number implies that t is one-one.

We finish the proof of the theorem by observing that the composition of one-one functions is one-one, and so the composition

$$t \circ (I \times I) \colon (0,1) \times (0,1) \to J \times J \to (0,1)$$

is one-one. The Schröder-Bernstein Theorem applies to give a one-one correspondence between $(0,1)$ and $(0,1) \times (0,1)$. Thus there is a one-one correspondence between $\mathbb{R}$ and $\mathbb{R} \times \mathbb{R}$. $\square$

Corollary 1.11. *There is a one-one correspondence between $\mathbb{R}^m$ and $\mathbb{R}^n$ for all positive integers m and n.*

The corollary follows by replacing $\mathbb{R}^2$ by $\mathbb{R}$ until $n = m$. A one-one correspondence is a relabelling of sets, and so as collections of labels we cannot distinguish between $\mathbb{R}^n$ and $\mathbb{R}^m$. It follows that a function $\mathbb{R}^m \to \mathbb{R}$ could be replaced by a function $\mathbb{R} \to \mathbb{R}$ by composing with the one-one correspondence $\mathbb{R} \to \mathbb{R}^m$. A function expressing the dependence of a physical quantity on two variables could be replaced by a function that depends on only one variable. This observation calls into question the dependence on a certain number of variables as a physically meaningful notion—perhaps such a dependence can always be reduced to fewer variables by this mathematical slight-of-hand. In the epigraph, Cantor expressed his surprise in his proof of Theorem 1.10, not in the result.

If we introduce more structure into the discussion, the notion of dimension emerges. For example, from the point of view of linear algebra, where we use the linear structure on $\mathbb{R}^m$ and $\mathbb{R}^n$ as vector spaces, we can distinguish between these sets by their linear dimension, the number of vectors in a basis.

If we apply the calculus to compare $\mathbb{R}^n$ and $\mathbb{R}^m$, we can ask if there exists a differentiable function $f \colon \mathbb{R}^n \to \mathbb{R}^m$ with an inverse that is also differentiable. At a given point of the domain, the derivative of such a differentiable mapping is a linear mapping, and the existence of a differentiable inverse implies that this linear mapping is invertible. Thus, by linear algebra, we deduce that $n = m$.

Between the realm of sets and the realm of the calculus lies the realm of topology—in particular, the study of continuous functions. The **main problem** addressed in this book is the following:

> **Suppose there is a continuous function** $f \colon \mathbb{R}^n \to \mathbb{R}^m$ **with a continuous inverse. Does this imply** $n = m$**?**

This problem is called the question of the topological *Invariance of Dimension*, and it was one of the principal problems faced by the mathematicians who first developed topology. The problem was important because the use of dimension in the description of the physical space we dwell in was called into question by Cantor's discovery. The first proof of the topological invariance of dimension used new methods of a combinatorial nature (Chapters 9, 10, 11).

The combinatorial aspects of topology play a similar role that approximation does in analysis: by approximating with manageable objects, we can manipulate the approximations fruitfully, sometimes identifying properties that are associated to the combinatorics, but which depend only on the topology of the limiting object. This approach was initiated by Poincaré and refined to a subtle tool by Brouwer. It was Brouwer who gave the first complete proof of the theorem of the topological invariance of dimension and his proof established the centrality of combinatorial approximation in the study of continuity.

Toward our goal of a proof of invariance of dimension, we begin by expanding the familiar definition of continuity to more general settings.

Exercises

1. Let $f \colon A \to B$ be any function with U, V subsets of B and X a subset of A. Prove the following about the preimage operation:
   a) $U \subset V$ implies $f^{-1}(U) \subset f^{-1}(V)$.
   b) $f^{-1}(U \cup V) = f^{-1}(U) \cup f^{-1}(V)$.
   c) $f^{-1}(U \cap V) = f^{-1}(U) \cap f^{-1}(V)$.
   d) $f(f^{-1}(U)) \subset U$.

    e) $f^{-1}(f(X)) \supset X$.

    f) If for any $U \subset B, f(f^{-1}(U)) = U$, then f is onto.

    g) If for any $X \subset A, f^{-1}(f(X)) = X$, then f is one-one.

2. Show that a set S and its power set $\mathcal{P}(S)$ cannot have the same cardinality. (Hints to an unexpected proof: Suppose there is an onto function $j \colon S \to \mathcal{P}(S)$. Define the subset of S,

$$T = \{s \in S \mid s \notin j(s)\} \in \mathcal{P}(S).$$

If j is surjective, then there is an element $t \in S$ with $j(t) = T$. Is $t \in T$?) Show that $\mathcal{P}(S)$ can be put in one-one correspondence with the set $\mathrm{map}(S, \{0, 1\})$ of functions from the set S to $\{0, 1\}$.

3. On the power set of a set X, $\mathcal{P}(X) = \{\text{subsets of } X\}$, we have the equivalence relation $U \cong V$ whenever there is a one-one correspondence between U and V. There is also a binary operation on $\mathcal{P}(X)$ given by taking unions:

$$\cup \colon \mathcal{P}(X) \times \mathcal{P}(X) \to \mathcal{P}(X), \quad \cup(U, V) = U \cup V,$$

where $U \cup V$ is the union of the subsets U and V. Show by example that the equivalence relation $\cong$ is not a congruence relation.

4. An equivalence relation, called the *equivalence kernel*, can be constructed from a function $f \colon A \to B$. The relation is on A and is defined by

$$x \sim y \Longleftrightarrow f(x) = f(y).$$

Show that this is an equivalence relation. Determine the relation that arises on $\mathbb{R}$ from the mapping $f(r) = \cos 2\pi r$. What equivalence kernel results from taking the canonical mapping $A \to [A]'$, where $\sim'$ is some equivalence relation on A?

Chapter 2

Metric and Topological Spaces

*Topology begins where sets are implemented with
some cohesive properties enabling one to define
continuity.*

SOLOMON LEFSCHETZ

In order to forge a *language of continuity*, we begin with familiar examples. Recall from single-variable calculus that a function $f \colon \mathbb{R} \to \mathbb{R}$ is *continuous at a point* $x_0 \in \mathbb{R}$ if for every $\epsilon > 0$, there is a $\delta > 0$ so that, whenever $|x - x_0| < \delta$, we have $|f(x) - f(x_0)| < \epsilon$. The route to generalization begins with the distance notion on the real line: the distance between the real numbers x and y is given by $|x - y|$. The general properties of a distance are abstracted in the the notion of a *metric space*, first introduced by MAURICE FRÉCHET (1878–1973) and named by Hausdorff.

Definition 2.1. A **metric space** is a set X together with a distance function $d \colon X \times X \to \mathbb{R}$ satisfying

i) $d(x, y) \geq 0$ for all $x, y \in X$ and $d(x, y) = 0$ if and only if $x = y$.

ii) $d(x, y) = d(y, x)$ for all $x, y \in X$.

iii) The Triangle Inequality: $d(x,y) + d(y,z) \geq d(x,z)$ for all $x, y, z \in X$.

The **open ball** of radius $\epsilon > 0$ centered at a point x in a metric space (X, d) is given by

$$B(x, \epsilon) = \{y \in X \mid d(x, y) < \epsilon\},$$

that is, the points in X within ϵ in distance from x.

The intuitive notion of 'near' can be made precise in a metric space: a point y is 'near' the point x if it is in $B(x, \epsilon)$ for ϵ suitably small.

Examples. (1) The most familiar example is $\mathbb{R}^n$. If $\mathbf{x} = (x_1, \ldots, x_n)$ and $\mathbf{y} = (y_1, \ldots, y_n)$, then the Euclidean metric is given by

$$d(\mathbf{x}, \mathbf{y}) = \|\mathbf{x} - \mathbf{y}\| = \sqrt{(x_1 - y_1)^2 + \cdots + (x_n - y_n)^2}.$$

In fact, one can endow $\mathbb{R}^n$ with other metrics, for example,

$$d_1(\mathbf{x}, \mathbf{y}) = \max\{|x_1 - y_1|, \ldots, |x_n - y_n|\}.$$

The nonnegative, nondegenerate, and symmetric conditions are clear for d_1. The triangle inequality follows in the same way as the proof in the next example.

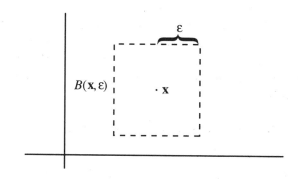

Notice that an open ball with this metric is an 'open box' as pictured here in $\mathbb{R}^2$.

(2) Let $X = \mathrm{Bdd}([0, 1], \mathbb{R})$ denote the set of *bounded functions* $f \colon [0, 1] \to \mathbb{R}$, that is, functions f for which there is a real number

$M(f)$ such that $|f(t)| < M(f)$ for all $t \in [0, 1]$. Define the distance between two such functions to be

$$d(f, g) = \mathrm{lub}_{t \in [0,1]}\{|f(t) - g(t)|\}.$$

Certainly $d(f, g) \geq 0$ and $d(f, g) = 0$ if and only if $f = g$. Furthermore, $d(f, g) = d(g, f)$. The triangle inequality is more subtle:

$$\begin{aligned}
d(f, h) &= \mathrm{lub}_{t \in [0,1]}\{|f(t) - h(t)|\} \\
&\leq \mathrm{lub}_{t \in [0,1]}\{|f(t) - g(t)| + |g(t) - h(t)|\} \\
&\leq \mathrm{lub}_{t \in [0,1]}\{|f(t) - g(t)|\} + \mathrm{lub}_{t \in [0,1]}\{|g(t) - h(t)|\} \\
&= d(f, g) + d(g, h).
\end{aligned}$$

An open ball in this metric space, $B(f, \epsilon)$, consists of all functions defined on $[0, 1]$ with graph in the stripe pictured:

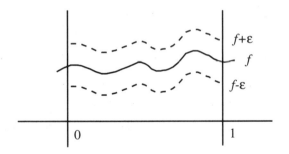

(3) Let X be any set and define

$$d(x, y) = \begin{cases} 0, & \text{if } x = y, \\ 1, & \text{if } x \neq y. \end{cases}$$

This is a perfectly good distance function—open balls are funny, however—either they consist of one point or the whole space depending on whether $\epsilon \leq 1$ or $\epsilon > 1$. The resulting metric space is called the *discrete metric space*.

Using open balls, we can rewrite the definition of a continuous real-valued function $f \colon \mathbb{R} \to \mathbb{R}$ to say:

> *A function $f \colon \mathbb{R} \to \mathbb{R}$ is continuous at $x_0 \in \mathbb{R}$ if for any $\epsilon > 0$, there is a $\delta > 0$ so that $B(x_0, \delta) \subset f^{-1}(B(f(x_0), \epsilon))$.*

The step from this definition of continuity to a general definition of continuous mappings of metric spaces is clear.

Definition 2.2. Suppose (X, d_X) and (Y, d_Y) are two metric spaces and $f\colon X \to Y$ is a function. Then f is **continuous at** $x_0 \in X$ if for any $\epsilon > 0$, there is a $\delta > 0$ so that $B(x_0, \delta) \subset f^{-1}(B(f(x_0), \epsilon))$. The function f is **continuous** if it is continuous at x_0 for all $x_0 \in X$.

For example, if $X = Y = \mathbb{R}^n$ with the usual Euclidean metric $d(\mathbf{x}, \mathbf{y}) = \|\mathbf{x} - \mathbf{y}\|$, then $f\colon \mathbb{R}^n \to \mathbb{R}^n$ is continuous at $\mathbf{x}_0$ if for any $\epsilon > 0$, there is a $\delta > 0$ so that whenever $\mathbf{x} \in B(\mathbf{x}_0, \delta)$, that is, $\|\mathbf{x} - \mathbf{x}_0\| < \delta$, then $\mathbf{x} \in f^{-1}(B(f(\mathbf{x}_0), \epsilon))$, which is to say, $f(\mathbf{x}) \in B(f(\mathbf{x}_0), \epsilon)$, or $\|f(\mathbf{x}) - f(\mathbf{x}_0)\| < \epsilon$. Thus we recover the ϵ–δ definition of continuity. We develop the generalization further.

Definition 2.3. A subset U of a metric space (X, d) is **open** if for any $u \in U$, there is an $\epsilon > 0$ so that $B(u, \epsilon) \subset U$.

We note the following properties of open subsets of metric spaces.

1) An open ball $B(x, \epsilon)$ is an open set in (X, d).

2) An arbitrary union of open subsets in a metric space is open.

3) The finite intersection of open subsets in a metric space is open.

Suppose $y \in B(x, \epsilon)$. Let $\delta = \epsilon - d(x, y) > 0$. Consider the open ball $B(y, \delta)$. If $z \in B(y, \delta)$, then $d(z, y) < \delta = \epsilon - d(x, y)$, or $d(z, y) + d(y, x) < \epsilon$. By the triangle inequality $d(z, x) \leq d(z, y) + d(y, x)$ and so $d(z, x) < \epsilon$ and $B(y, \delta) \subset B(x, \epsilon)$. Thus $B(x, \epsilon)$ is open.

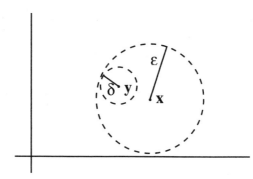

Suppose $\{U_\alpha, \ \alpha \in I\}$ is a collection of open subsets of X. If $x \in \bigcup_{\alpha \in I} U_\alpha$, then $x \in U_\beta$ for some $\beta \in I$. But U_β is open so there is an $\epsilon > 0$ with $B(x, \epsilon) \subset U_\beta \subset \bigcup_{\alpha \in I} U_\alpha$. Therefore, the union $\bigcup_{\alpha \in I} U_\alpha$ is open.

Suppose $U_1, U_2, \ldots, U_n$ are open in X, and suppose $x \in U_1 \cap U_2 \cap \cdots \cap U_n$. Then $x \in U_i$ for $i = 1, 2, \ldots, n$ and since each U_i is open there are $\epsilon_1, \epsilon_2, \ldots, \epsilon_n > 0$ with $B(x, \epsilon_i) \subset U_i$. Let $\epsilon = \min\{\epsilon_1, \epsilon_2, \ldots, \epsilon_n\}$. Then $\epsilon > 0$ and $B(x, \epsilon) \subset B(x, \epsilon_i) \subset U_i$ for all i, so $B(x, \epsilon) \subset U_1 \cap \cdots \cap U_n$ and the intersection is open.

We can use the language of open sets to rephrase the definition of continuity for metric spaces.

Theorem 2.4. *A function $f \colon X \to Y$ between metric spaces (X, d) and (Y, d) is continuous if and only if for any open subset V of Y, the subset $f^{-1}(V)$ is open in X.*

Proof. Suppose $x_0 \in X$ and $\epsilon > 0$. Then $B(f(x_0), \epsilon)$ is an open set in Y. By assumption, $f^{-1}(B(f(x_0), \epsilon))$ is an open subset of X. Since $x_0 \in f^{-1}(B(f(x_0), \epsilon))$, there is a $\delta > 0$ with $B(x_0, \delta) \subset f^{-1}(B(f(x_0), \epsilon))$ and so f is continuous at x_0.

Suppose that V is an open set in Y, and that $x \in f^{-1}(V)$. Then $f(x) \in V$ and there is an $\epsilon > 0$ with $B(f(x), \epsilon) \subset V$. Since f is continuous at x, there is a $\delta > 0$ with $B(x, \delta) \subset f^{-1}(B(f(x), \epsilon)) \subset f^{-1}(V)$. Thus, for each $x \in f^{-1}(V)$, there is a $\delta > 0$ with $B(x, \delta) \subset f^{-1}(V)$, that is, $f^{-1}(V)$ is open in X. $\qquad \square$

It follows from this theorem that, for metric spaces, continuity may be described entirely in terms of open sets. To study continuity in general we take the next step and focus on the collection of open sets. The key features of the structure of open sets in metric spaces may be abstracted to the following definition, first given by Hausdorff in 1914 [**33**].

Definition 2.5. Let X be a set and $\mathcal{T}$ a collection of subsets of X called **open sets**. The collection $\mathcal{T}$ is called a **topology** on X if

1) we have that $\emptyset \in \mathcal{T}$ and $X \in \mathcal{T}$,

    2) the union of an arbitrary collection of members of $\mathcal{T}$ is in $\mathcal{T}$,

    3) the finite intersection of members of $\mathcal{T}$ is in $\mathcal{T}$.

The pair $(X, \mathcal{T})$ is called a **(topological) space**.

It is important to note that open sets are basic and determine the topology. Open set does not always refer to the 'open' sets we are used to in $\mathbb{R}^n$. Let's consider some examples.

Examples. (1) If (X, d) is a metric space, we defined a subset U of X to be open if for any $x \in U$, there is an $\epsilon > 0$ with $B(x, \epsilon) \subset U$, as above. This collection of open sets defines a topology on X called the *metric topology.*

(2) For any set X, let $\mathcal{T}_1 = \{X, \emptyset\}$. This collection trivially satisfies the criteria for being a topology and is called the **indiscrete topology** on X. Let $\mathcal{T}_2 = \mathcal{P}(X)$ be the set of all subsets of X. This collection trivially satisfies the conditions to be a topology and is called the **discrete topology** on X. It has the same open sets as the metric topology in X with the discrete metric. It is the largest topology possible on a set (the most open sets), while the indiscrete topology is the smallest topology.

(3) For the set with only two elements $X = \{0, 1\}$ consider the collection of open sets given by $\mathcal{T}_S = \{\emptyset, \{0\}, \{0, 1\}\}$. The reader can quickly check that $\mathcal{T}_S$ is a topology. This topological space is called the *Sierpinski 2-point space*.

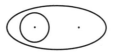

(4) Let X be an infinite set. Define $\mathcal{T}_{FC} = \{U \subset X \mid U = \emptyset$ or $X - U$ is finite$\}$. We show that $\mathcal{T}_{FC}$ is a topology:

    1) The empty set is already in $\mathcal{T}_{FC}$; X is open since $X - X = \emptyset$, which is finite.

    2) If $\{U_\alpha, \alpha \in J\}$ is an arbitrary collection of open sets, then

$$X - \bigcup_{\alpha \in J} U_\alpha = \bigcap_{\alpha \in J} (X - U_\alpha)$$

by DeMorgan's Law. Each $X - U_\alpha$ is finite or all of X so we have $X - \bigcup_{\alpha \in J} U_\alpha$ is finite or all of X and so $\bigcup_{\alpha \in J} U_\alpha$ is open.

3) If $U_1, U_2, \ldots, U_n$ are open, then $X - (U_1 \cap \cdots \cap U_n) = (X - U_1) \cup \cdots \cup (X - U_n)$, again by DeMorgan's Law. Either one gets all of X or a finite union of finite sets and so an open set.

The collection $\mathcal{T}_{FC}$ is called the *finite-complement topology* on the infinite set X. The finite-complement topology will offer an example later of how strange convergence properties can become in some topological spaces.

(5) On a three-point set there are nine distinct topologies, where by distinct we mean up to renaming the points. The distinct topologies are shown in the following diagram.

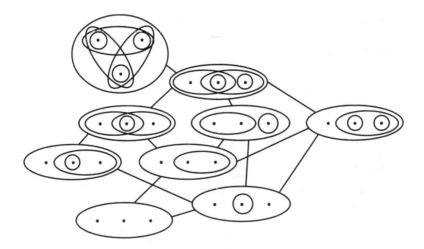

Given two topologies $\mathcal{T}, \mathcal{T}'$ on a given set X we say $\mathcal{T}$ is **finer** than $\mathcal{T}'$ if $\mathcal{T}' \subset \mathcal{T}$. Equivalently we say $\mathcal{T}'$ is **coarser** than $\mathcal{T}$. For example, on any set the indiscrete topology is coarser and the discrete topology is finer than any other topology. The finite-complement topology on $\mathbb{R}$ is strictly coarser than the metric topology. I have added a line joining comparable topologies in the diagram of the distinct topologies on a three-point set. Coarser is lower in this case, and the relation is transitive. As we will see later, the ordering of topologies plays a role in the continuity of functions.

On a given set X it would be nice to have a way of generating topologies. One way is to use a *basis* for the topology:

Definition 2.6. A collection of subsets, $\mathcal{B}$, of a set X is a **basis for a topology on** X if

1) for all $x \in X$, there is a $B \in \mathcal{B}$ with $x \in B$,

2) $x \in B_1 \in \mathcal{B}$ and $x \in B_2 \in \mathcal{B}$, then there is some $B_3 \in \mathcal{B}$ with $x \in B_3 \subset B_1 \cap B_2$.

Proposition 2.7. *If $\mathcal{B}$ is a basis for a topology on a set X, then the collection of subsets*

$$\mathcal{T}_\mathcal{B} = \{ \textstyle\bigcup_{\alpha \in A} B_\alpha \mid A \text{ is any index set and } B_\alpha \in \mathcal{B} \text{ for all } \alpha \in A \}$$

*is a topology on X called the **topology generated by the basis** $\mathcal{B}$.*

Proof. We show that $\mathcal{T}_\mathcal{B}$ satisfies the axioms for a topology. By the definition of a basis, we can write $X = \bigcup_{B \in \mathcal{B}} B$ and $\emptyset = \bigcup_{i \in \emptyset} U_i$; so X and $\emptyset$ are in $\mathcal{T}_\mathcal{B}$. If U_j, $j \in J$, are in $\mathcal{T}_\mathcal{B}$, then write $U_j = \bigcup_{\alpha \in A_j} B_\alpha$ and so

$$\bigcup_{j \in J} U_j = \bigcup_{j \in J} \left(\bigcup_{\alpha \in A_j} B_\alpha \right) = \bigcup_{\alpha \in \bigcup_{j \in J} A_j} B_\alpha$$

and so $\mathcal{T}_\mathcal{B}$ is closed under arbitrary unions.

For finite intersections we prove the case of two sets and apply induction. As above

$$U \cap V = \left(\bigcup_{\alpha \in A} B_\alpha \right) \cap \left(\bigcup_{\gamma \in C} B_\gamma \right).$$

If $x \in U \cap V$, then $x \in B_{\alpha_1} \cap B_{\gamma_1}$ for some $\alpha_1 \in A$ and $\gamma_1 \in C$ and so there is a B_3^x in $\mathcal{B}$ with $x \in B_3^x \subset B_{\alpha_1} \cap B_{\gamma_1} \subset U \cap V$. We obtain such a set B_3^x for each x in $U \cap V$ and so we deduce

$$U \cap V \subset \bigcup_{x \in U \cap V} B_3^x \subset U \cap V.$$

Since we have written $U \cap V$ as a union of basis sets, $U \cap V$ is in $\mathcal{T}_\mathcal{B}$. $\qquad \square$

Examples. (1) The basis $\mathcal{B} = \{X\}$ generates the indiscrete topology, while $\mathcal{B} = \{\{x\} \mid x \in X\}$ generates the discrete topology.

(2) On $\mathbb{R}$, we can take the family of subsets $\mathcal{B} = \{(a,b) \mid a < b\}$. This is a basis since $(a,b) \cap (c,d)$ is one of (a,b), (a,d), (c,b), or (c,d). This leads to the metric topology on $\mathbb{R}$. In fact, we can take a smaller set

$$\mathcal{B}_u = \{(a,b) \mid a < b \text{ and } a,b \text{ rational numbers}\}.$$

For any (r,s) with $r, s \in \mathbb{R}$ and $r < s$, we can write $(r,s) = \bigcup (a,b)$ for $r < a < b < s$ and $a, b \in \mathbb{Q}$. Thus $\mathcal{B}_u$ also generates the usual metric topology, but $\mathcal{B}_u$ is a countable set. We say that a space is **second countable** when it has a basis for its topology that is countable as a set.

(3) More generally, if (X,d) is a metric space, then the collection

$$\mathcal{B}_d = \{B(x,\epsilon) \mid x \in X, \epsilon > 0\}$$

is a basis for the metric topology in X. We check the intersection condition: Suppose $z \in B(x,\epsilon)$ and $z \in B(y,\epsilon')$; then let $0 < \delta < \min\{\epsilon - d(x,z), \epsilon' - d(y,z)\}$. Consider $B(z,\delta)$ and suppose $w \in B(z,\delta)$. Then

$$d(x,w) \leq d(x,z) + d(z,w)$$
$$< d(x,z) + \delta \leq d(x,z) + \epsilon - d(x,z) = \epsilon.$$

Likewise, $d(y,w) < \epsilon'$ and so $B(z,\delta) \subset B(x,\epsilon) \cap B(y,\epsilon')$ as required.

(4) A nonstandard basis for a topology on $\mathbb{R}$ is given by $\mathcal{B}_{ho} = \{[a,b) \mid a < b\}$. This basis generates the *half-open topology* on $\mathbb{R}$. Notice that the half-open topology is *strictly* finer than the metric topology since

$$(a,b) = \bigcup_{n=k}^{\infty} [a + (1/n), b)$$

for k large enough that $a + (1/k) < b$. However, no subset $[a,b)$ is a union of open intervals.

Proposition 2.8. *If $\mathcal{B}_1$ and $\mathcal{B}_2$ are bases for topologies on a set X, and for all $x \in X$ and $x \in B_1 \in \mathcal{B}_1$, there is a B_2 with $x \in B_2 \subseteq B_1$ and $B_2 \in \mathcal{B}_2$, then $\mathcal{T}_{\mathcal{B}_2}$ is finer than $\mathcal{T}_{\mathcal{B}_1}$.*

The proof is left as an exercise. The proposition applies to metric spaces. Given two metrics on a space, when do they give the same topology? Let d_1 and d_2 denote the metrics and $B_1(x, \epsilon)$, $B_2(x, \epsilon)$ the open balls of radius ϵ at x given by each metric, respectively. The proposition is satisfied if, for $i = 2$, $j = 1$ and again for $i = 1$, $j = 2$, for any $y \in B_i(x, \epsilon)$, there is an $\epsilon' > 0$ with $B_j(y, \epsilon') \subset B_i(x, \epsilon)$. Then the topologies are equivalent. For example, the two metrics defined on $\mathbb{R}^m$,

$$d_1(\mathbf{x}, \mathbf{y}) = \sqrt{(x_1 - y_1)^2 + \cdots + (x_m - y_m)^2},$$
$$d_2(\mathbf{x}, \mathbf{y}) = \max\{|x_i - y_i| \mid i = 1, \ldots, m\},$$

give the same topology.

Continuity

Having identified the places where continuity can happen, namely, topological spaces, we define what it means to be a continuous function between spaces.

Definition 2.9. Let $(X, \mathcal{T})$ and $(Y, \mathcal{T}')$ be topological spaces and $f \colon X \to Y$ a function. We say that f is **continuous** if whenever V is open in Y, $f^{-1}(V)$ is open in X.

This simple definition generalizes the definition of continuous function between metric spaces, and hence recovers our classical definition from the calculus.

The identity mapping $\mathrm{id} \colon (X, \mathcal{T}) \to (X, \mathcal{T})$ is always continuous. However, if we change the topology on the domain or codomain, this may not be true. For example, $\mathrm{id} \colon (\mathbb{R}, \text{usual}) \to (\mathbb{R}, \text{half-open})$ is not continuous since $\mathrm{id}^{-1}([0, 1)) = [0, 1)$, which is not open in the usual topology. The following proposition is an easy observation.

Proposition 2.10. *If $\mathcal{T}$ and $\mathcal{T}'$ are topologies on a set X, then the identity mapping $\mathrm{id} \colon (X, \mathcal{T}) \to (X, \mathcal{T}')$ is continuous if and only if $\mathcal{T}$ is finer than $\mathcal{T}'$.*

With this formulation of continuity it is straightforward to give proofs of some of the properties of continuous functions.

Theorem 2.11. *Given two continuous functions $f \colon X \to Y$ and $g \colon Y \to Z$, the composite function $g \circ f \colon X \to Z$ is continuous.*

Proof. If V is open in Z, then $g^{-1}(V) = U$ is open in Y and so $f^{-1}(U)$ is open in X. But $(g \circ f)^{-1}(V) = f^{-1}(g^{-1}(V)) = f^{-1}(U)$, so $(g \circ f)^{-1}(V)$ is open in X and $g \circ f$ is continuous. $\square$

We next give a key definition for topology—the means of comparison of spaces.

Definition 2.12. A function $f \colon (X, \mathcal{T}_X) \to (Y, \mathcal{T}_Y)$ is a **homeomorphism** if f is continuous, one-one, onto, and has a continuous inverse. We say $(X, \mathcal{T}_X)$ and $(Y, \mathcal{T}_Y)$ are **homeomorphic topological spaces** if there is a homeomorphism $f : (X, \mathcal{T}_X) \to (Y, \mathcal{T}_Y)$. A property of a space $(X, \mathcal{T}_X)$ is said to be a **topological property** if, whenever $(Y, \mathcal{T}_Y)$ is homeomorphic to $(X, \mathcal{T}_X)$, then the space $(Y, \mathcal{T}_Y)$ also has the property.

Examples. (1) We may take all functions known from the calculus to be continuous functions as having been proved continuous in our language. For example, the mapping $\arctan \colon \mathbb{R} \to (-\pi/2, \pi/2)$ is a homeomorphism. Notice that the metric idea of a subset being of infinite extent is not a topological notion.

(2) By the definition of the indiscrete and discrete topologies, any function $f \colon (X, \text{discrete}) \to (Y, \mathcal{T})$ is continuous as is any function $g \colon (X, \mathcal{T}) \to (Y, \text{indiscrete})$. A partial order is obtained on topologies on a set X by $\mathcal{T} \le \mathcal{T}'$ if the identity mapping $\text{id} \colon (X, \mathcal{T}) \to (X, \mathcal{T}')$ is continuous. This order is the relation of fineness.

The definition of homeomorphism makes topology the geometry of topological properties in the sense of Klein's *Erlangen Program* [45]. We treat a *figure* as a subset of a space $(X, \mathcal{T})$ and the homeomorphisms $f \colon X \to X$ are the transformations carrying a figure to a "congruent" figure.

The simplest topological property is the cardinality of the space, because a homeomorphism is a one-one correspondence. A more topological example is the notion of second countability.

Proposition 2.13. *The property of being second countable is a topological property.*

Proof. Suppose $(X, \mathcal{T})$ has a countable basis $\{U_i, i = 1, 2, \ldots\}$. Suppose that $f \colon (X, \mathcal{T}_X) \to (Y, \mathcal{T}_Y)$ is a homeomorphism. Write $g = f^{-1} \colon (Y, \mathcal{T}_Y) \to (X, \mathcal{T}_X)$ for the inverse homeomorphism. Let $V_i = g^{-1}(U_i)$. Then the proposition follows from a proof that $\{V_i : i = 1, 2, \ldots\}$ is a countable basis for Y. To prove this we take any open set $W \subset Y$ and show for all $w \in W$ there is some j with $w \in V_j \subset W$. Let $O = f^{-1}(W)$ and $u = f^{-1}(w) = g(w)$ so that $u \in O \subset X$. Then there is some j with $u \in U_j \subset O$. Apply g^{-1} to get $w \in V_j = g^{-1}(U_j) \subset g^{-1}(O)$. But $g^{-1}(O) = W$ so $w \in V_j \subset W$ as desired, and $(Y, \mathcal{T}_Y)$ is second countable. $\square$

Later chapters will be devoted to some of the most important topological properties.

Exercises

1. Prove Proposition 2.8.

2. Another way to generate a topology on a set X is from a **subbasis**, which is a set $\mathcal{S}$ of subsets of X such that, for any $x \in X$, there is an element $S \in \mathcal{S}$ with $x \in S$. Show that the collection $\mathcal{B}_\mathcal{S} = \{S_1 \cap \cdots \cap S_n \mid S_i \in \mathcal{S}, n > 0\}$ is a basis for a topology on X. Show that the set $\{(-\infty, a), (b, \infty) \mid -\infty < a, b < \infty\}$ is a subbasis for the usual topology on $\mathbb{R}$.

3. Suppose that X is an *uncountable* set and that x_0 is some given point in X. Let $\mathcal{T}_F$ be the collection of subsets $\mathcal{T}_F = \{U \subset X \mid X - U \text{ is finite or } x_0 \notin U\}$. Show that $\mathcal{T}_F$ is a topology on X, called the **Fort topology**.

4. Suppose $X = \mathrm{Bdd}([0, 1], \mathbb{R})$ is the metric space of bounded real-valued functions on $[0, 1]$. Let $F \colon X \to \mathbb{R}$ be defined by $F(f) = f(1)$. Show that this is a continuous function when $\mathbb{R}$ has the usual topology.

5. A space $(X, \mathcal{T})$ is said to have the **fixed point property** (FPP) if any continuous function $f \colon (X, \mathcal{T}) \to (X, \mathcal{T})$ has a fixed point, that is, there is some $x \in X$ with $f(x) = x$. Show that the FPP is a topological property.

5. The *taxicab metric* on $\mathbb{R}^n$ is given by

$$d(\mathbf{x}, \mathbf{y}) = |x_1 - y_1| + \cdots + |x_n - y_n|.$$

Prove that this is indeed a metric on $\mathbb{R}^n$. Describe the open balls in the taxicab metric on $\mathbb{R}^2$. How do the usual topology and the taxicab metric topology compare on $\mathbb{R}^n$?

6. A space $(X, \mathcal{T})$ is said to be a T_1-space if for any $x \in X$, the complement of $\{x\}$ is open in X. Show that a metric space is T_1. Which of the topologies on the three-point set are T_1? Show that being T_1 is a topological property.

7. We displayed the nine distinct topologies on a three element set in this chapter. The sequence of integers

t_n = number of distinct topologies on a set of n elements

may be found in Neil Sloane's On-Line Encyclopedia of Integer Sequences with ID Number A001930. The first few values of t_n, beginning with t_0, are given by

$$1, 1, 3, 9, 33, 139, 718, 4535, 35979, 363083,$$
$$4717687, 79501654, 1744252509$$

See how far you can get finding the 33 distinct topologies on a set of four elements.

URL: http://www.research.att.com/projects/OEIS?Anum=A001930

Chapter 3

Geometric Notions

At the basis of the distance concept lies, for example, the concept of convergent point sequences and their defined limits, and one can, by choosing these ideas as those fundamental to point set theory, eliminate the notions of distance.

FELIX HAUSDORFF

By choosing open sets as the basic notion we can generalize familiar analytic and geometric notions from Euclidean space to the new setting of topology. Two fundamental notions were introduced by Cantor in his work [**13**] on analysis. In the language of topology, these ideas have simple definitions.

Definition 3.1. Let $(X, \mathcal{T})$ be a topological space. A subset K of X is **closed** if its complement in X is open. If $A \subseteq X$, where X is a topological space and $x \in X$, then x is a **limit point** of A, if, whenever $U \subset X$ is open and $x \in U$, there is some $y \in U \cap A$, with $y \neq x$.

Closed sets are the natural generalization of closed sets in $\mathbb{R}^n$. Notice that an arbitrary subset of a topological space can be neither open nor closed, for example, $[a, b) \subset \mathbb{R}$ in the usual topology. A slogan to remember is that *"a subset is not a door."*

In a metric space the notion of a limit point w of a subset A is given by a sequence $\{x_i, i = 1, 2, \ldots\}$ with $x_i \in A$ for all i and $\lim_{i \to \infty} x_i = w$. The limit is defined as usual: for any $\epsilon > 0$, there is an integer N for which whenever $n \geq N$, we have $d(x_n, w) < \epsilon$. We distinguish two cases: If $w \in A$, then we can choose the constant sequence to converge to w. For $w \notin A$ to be a limit point, we want, for each $\epsilon > 0$, that there be some other point $a_\epsilon \in A$ with $a_\epsilon \neq w$ and $a_\epsilon \in B(w, \epsilon)$. When w is a limit point of A, such points a_ϵ always exist. If we form the sequence $\{x_i = a_{1/i}\}$, then $\lim_{i \to \infty} x_i = w$ follows. Conversely, if there is a sequence of infinitely many distinct points $x_i \in A$ with $\lim_{i \to \infty} x_i = w$, then w is a limit point of A.

The limit points of a subset of a metric space are "near" the subset. In the most general topological spaces, the situation can be quite different. Consider $\mathbb{R}$ with the finite-complement topology and let $A = \mathbb{Z}$, the set of integers in $\mathbb{R}$. Choose any real number r and suppose U is an open set containing r. Then $U = \mathbb{R} - \{s_1, s_2, \ldots, s_k\}$ for some choices of real numbers $s_1, \ldots, s_k$. Since this set leaves out only finitely many points and $\mathbb{Z}$ is infinite, there are infinitely many integers in U and certainly one not equal to r. Thus r is a limit point of $\mathbb{Z}$. This is an extreme case—every point in the space is a limit point of a proper subset.

Closed sets and limit points are related.

Proposition 3.2. *A subset K of a topological space $(X, \mathcal{T})$ is closed if and only if it contains all of its limit points.*

Proof. Suppose K is closed, $x \in X$ is some point, and $x \notin K$. Then $x \in X - K$ and $X - K$ is open. So x is contained in an open set that does not intersect K, and therefore, x is not a limit point of K. Thus all limit points of K must be in K.

Suppose K contains all of its limit points. Let $x \in X - K$. Then x is not a limit point and so there exists an open set U^x with $x \in U^x$ and $U^x \cap K = \emptyset$, that is, $U^x \subset X - K$. Since we can find such an open set U^x for all $x \in X - K$, we have

$$X - K \subset \bigcup_{x \in X - K} U^x \subset X - K.$$

We have written $X - K$ as a union of open sets. Hence $X - K$ is open and K is closed. ☐

Let $(X, \mathcal{T})$ be a topological space and A an arbitrary subset of X. We associate to A subsets definable with the open sets in the topology as follows:

Definition 3.3. The **interior** of A is the largest open set contained in A, that is,

$$\text{int } A = \bigcup_{U \subseteq A, \text{open}} U.$$

The **closure** of A is the smallest closed set in X containing A, that is,

$$\text{cls } A = \bigcap_{K \supseteq A, \text{closed}} K.$$

These operations tell us something geometric about subsets, for example, the subset $\mathbb{Q} \subset (\mathbb{R}, \text{usual})$ has empty interior and closure all of $\mathbb{R}$. To see this suppose $U \subset \mathbb{R}$ is open. Then there is an interval $(a, b) \subset U$ for some $a < b$. Since (a, b) contains an irrational number, $(a, b) \cap \mathbb{R} - \mathbb{Q} \neq \emptyset$, $U \not\subset \mathbb{Q}$ and so $\text{int } \mathbb{Q} = \emptyset$. If $\mathbb{Q} \subset K$ is a closed subset of $\mathbb{R}$, then $\mathbb{R} - K$ is open and contains no rationals. It follows that it contains no interval because every nonempty interval of real numbers contains a rational number. Thus $\mathbb{R} - K = \emptyset$ and $\text{cls } \mathbb{Q} = \mathbb{R}$.

The operation of closure ought to be a kind of 'closing' up of the set by putting in all the 'ragged edges.' We make this precise as follows:

Proposition 3.4. *If $A \subset X$, where X is a topological space, then* $\text{cls } A = A \cup A'$, *where*

$$A' = \{\text{limit points of } A\}.$$

*A' is called the **derived set** of A.*

Proof. By definition, $\text{cls } A$ is closed and contains A so $A \subset \text{cls } A$. It follows that if $x \notin \text{cls } A$, then there exists an open set U containing x with $U \cap A = \emptyset$ and so $x \notin A$ and $x \notin A'$. This shows $A \cup A' \subset \text{cls } A$. To show the other containment, suppose $y \in \text{cls } A$ and V is an open set containing y. If $V \cap A = \emptyset$, then $A \subset (X - V)$, which is a closed set, and so $\text{cls } A \subset (X - V)$. But then $y \notin \text{cls } A$, a contradiction. If

$y \in \operatorname{cls} A$ and $y \notin A$, then, for any open set V with $y \in V$, we have $V \cap A \neq \emptyset$ and so y is a limit point of A. Thus $\operatorname{cls} A \subset A \cup A'$.   □

For any subset $A \subset X$, we have the following sequence of subsets:

$$\operatorname{int} A \subset A \subset \operatorname{cls} A = A \cup A'.$$

We add another more refined distinction between points in the closure.

Definition 3.5. Let A be a subset of X, a topological space. A point $x \in X$ is in the **boundary** of A, if for any open set $U \subset X$ with $x \in U$, we have $U \cap A \neq \emptyset$ and $U \cap (X - A) \neq \emptyset$. The set of points in the boundary of A is denoted $\operatorname{bdy} A$.

A boundary point of a subset is "on the edge" of the set. For example, suppose $A = (0,1] \cup \{2\}$ in $\mathbb{R}$ with the usual topology. The point 0 is a boundary point and a point in the derived set, but not in A; 1 is a boundary point, a point in the derived set, and a point in A; and 2 is boundary point, not in the derived set, but in A.

The boundary points lie outside the interior of A. We next see how the boundary relates to the closure.

Proposition 3.6. $\operatorname{cls} A = \operatorname{int} A \cup \operatorname{bdy} A$.

Proof. Suppose $x \in \operatorname{bdy} A$ and $K \subset X$ is closed with $A \subset K$. If $x \notin K$, then the open set $V = X - K$ contains x. Since $x \in \operatorname{bdy} A$, we have $V \cap A \neq \emptyset \neq V \cap (X - A)$. But $A \subset K$ implies $V \cap A = \emptyset$, a contradiction. Thus $\operatorname{bdy} A \subset \operatorname{cls} A$, and so $\operatorname{bdy} A \cup \operatorname{int} A \subset \operatorname{cls} A$.

We have already shown that $A \cup A' = \operatorname{cls} A$. If $x \in A - \operatorname{int} A$, then for any open set U containing x, $U \cap (X - A) \neq \emptyset$, otherwise x would be in the interior of A. By virtue of $x \in A$, $U \cap A \neq \emptyset$, so $x \in \operatorname{bdy} A$. Thus $\operatorname{int} A \cup \operatorname{bdy} A \supset A$. Consider $y \in A' \cap (X - A)$ and any open set V containing y. Since $y \in A'$, $V \cap A \neq \emptyset$. Also $V \cap (X - A) \neq \emptyset$ since $y \notin A$. Thus A' is a subset of $\operatorname{bdy} A$ and $\operatorname{cls} A \subset \operatorname{int} A \cup \operatorname{bdy} A$.   □

In a metric space, the notion of limit point agrees with the natural idea of the limit of a sequence of points from the subset. We next generalize convergence to topological spaces.

Definition 3.7. A sequence $\{x_n\}$ of points in a topological space $(X, \mathcal{T})$ is said to **converge to a point** $x \in X$, if for any open set U containing x, there is a positive integer $N = N(U)$ so that $x_n \in U$ whenever $n \geq N$.

This definition includes the notion of convergence in a metric space. In a general topological space, convergence of a sequence can be very strange. For example, consider the following topology on a nonempty set X: Let $x_0 \in X$ be chosen once and for all. Define $\mathcal{T}_{IP} = \{\emptyset \text{ or } U \subset X \text{ with } x_0 \in U\}$. This set of subsets determines a topology on X called the **included point topology**. (Check for yourself that $\mathcal{T}_{IP}$ is a topology.) Suppose $\{x_n\}$ is the constant sequence of points $x_n = x_0$ for all n. The sequence converges to $y \in X$, for *any* y: Any open set containing y, being nonempty, contains x_0. Thus *a constant sequence converges to every other point in the space* $(X, \mathcal{T}_{IP})$.

This case is extreme and it shows how wild an example a generalization can produce. Some further conditions keep such pathology in check. For example, to guarantee that a constant sequence converges only to the given point (and not other points as well), one needs at least one open set away from the point. The condition "X is a T_1-space", introduced in the previous exercises, requires that singleton sets be closed. A constant sequence can converge only to itself because there is an open set separating other points from it. We next introduce another formulation of the T_1 condition, placing it in a family of such conditions.

Definition 3.8. A topological space X is said to satisfy the T_1 **axiom** (Trennungsaxiom) if given two points $x, y \in X$, there are open sets U, V with $x \in U$, $y \notin U$ and $y \in V$, $x \notin V$. A topological space is said to satisfy the **Hausdorff condition** if given two points $x, y \in X$ there are open sets U, V with $x \in U$, $y \in V$, and $U \cap V = \emptyset$. The Hausdorff condition is also called the T_2 **axiom**.

Proposition 3.9. *A space X satisfies the T_1 axiom if and only if a finite subset of points in X is closed.*

Proof. Since a finite union of closed sets is closed, it suffices to check only a singleton subset. Suppose $x \in X$ and X is T_1; we show that $\{x\}$ is closed. Let y be in X, $y \neq x$. Then, by the T_1 axiom, there is an open set with $y \in U$, $x \notin U$. Denote this set by U_y. We have $U_y \subset X - \{x\}$. This can be done for each point $y \in X - \{x\}$ and we get

$$X - \{x\} \subset \bigcup_{y \in X - \{x\}} U_y \subset X - \{x\}.$$

Thus $X - \{x\}$ is a union of open sets, and $\{x\}$ is closed.

Conversely, suppose every singleton subset is closed in X. If x, $y \in X$ with $x \neq y$, then $x \in X - \{y\}$, $y \notin X - \{y\}$, and $X - \{y\}$ is open in X. Similarly, $y \in X - \{x\}$ and $x \notin X - \{x\}$, an open set in X. □

The T_1 axiom excludes some strange convergence behavior, but it is not enough to guarantee the uniqueness of limits. For example, if $(X, \mathcal{T}) = (\mathbb{R}, \mathcal{T}_{FC})$, the finite-complement topology on $\mathbb{R}$, then the T_1 axiom holds but the sequence of positive integers $\{1, 2, 3, \ldots\}$ converges to every real number. The Hausdorff condition remedies this pathology.

Theorem 3.10. *In a Hausdorff space, the limit of a sequence is unique.*

Proof. Suppose $\{x_n\}$ converges to x and to y with $x \neq y$. By the Hausdorff condition there are open sets U, V with $x \in U$, $y \in V$ such that $U \cap V = \emptyset$. But the definition of convergence gives integers $N = N(U)$ and $M = M(V)$ with $x_n \in U$ for $n \geq N$ and $x_m \in V$ for $m \geq M$. Take $L = \max\{N, M\}$; then $x_\ell \in U \cap V$ for $\ell \geq L$. But this cannot be, because $U \cap V = \emptyset$, so our assumption $x \neq y$ fails. □

An infinite set with the finite-complement topology is not Hausdorff.

A nice feature of the space $(\mathbb{R}, \text{usual})$ is its countable basis; thus open sets are expressible in a nice way. Another remarkable feature of $\mathbb{R}$ is the manner in which $\mathbb{Q}$ sits in $\mathbb{R}$. In particular, $\text{cls}\,\mathbb{Q} = \mathbb{R}$. We identify these features in the general setting of topological spaces.

Definition 3.11. A subset A of a topological space X is **dense** if $\text{cls}\,A = X$. A topological space is **separable** (or Fréchet) if it has a countable dense subset.

Theorem 3.12. *A separable metric space is second countable.*

Proof. Suppose A is a countable dense subset of (X, d). Consider the collection of open balls

$$\{B(a, p/q) \mid a \in A, p/q > 0, p/q \in \mathbb{Q}\}.$$

If U is an open set in X and $x \in U$, then there is an $\epsilon > 0$ with $B(x, \epsilon) \subset U$. Since $\operatorname{cls} A = X$, there is a point $a \in A \cap B(x, \epsilon/2)$. Consider $B(a, p/q)$ where p/q is rational and $d(a, x) < p/q < \epsilon/2$. Then $x \in B(a, p/q) \subset B(x, \epsilon) \subset U$. Repeat this procedure for each $x \in U$ to show $U \subset \bigcup_a B(a, p/q) \subset U$ and this collection of open balls is a basis for the topology on X. The collection is countable since a countable union of countable sets is countable. $\qquad \square$

The theorem applies to $(\mathbb{R}, \text{usual})$ and $\mathbb{Q} \subset \mathbb{R}$. Let $C^\infty([0,1], \mathbb{R})$ denote the set of all smooth functions $[0, 1] \to \mathbb{R}$, that is, functions possessing continuous derivatives of every order. From real analysis we know that any smooth function on $[0, 1]$ is bounded (a proof of this appears in Chapter 6) and so we can equip $C^\infty([0, 1], \mathbb{R})$ with the metric $d(f, g) = \max_{t \in [0,1]}\{|f(t) - g(t)|\}$. The Stone-Weierstrass theorem ([**70**]) implies that the countable set of polynomials with rational coefficients is dense in the metric space $(C^\infty([0, 1], \mathbb{R}), d)$. The proof follows by taking Taylor polynomials and approximating the coefficients by rationals. Thus $C^\infty([0, 1], \mathbb{R})$ is second countable.

When we defined continuity of a function in the calculus, we first define what it means to be continuous at a point. This is a *local* notion that requires only information about the behavior of the function close to the point. To be continuous in the calculus, a function must be continuous at every point of its domain, and this is a *global* condition. The topological formulation of continuous is global, though it can be made local to a point. Many properties of spaces have a local variant that expresses dependence on a chosen point. For example, we give a local version of second countability.

Definition 3.13. A topological space is **first countable** if for each $x \in X$ there is a collection of open sets $\{U_i^x \mid i = 1, 2, 3, \dots\}$ such that, for any V open in X with $x \in V$, there is one of these open sets U_j^x with $x \in U_j^x \subset V$.

A metric space is first countable taking the open balls centered at a point with rational radius for the collection U_i^x. The corresponding **global** condition is a countable basis for the entire space, that is, second countability.

The condition of first countability allows us to formulate the notion of limit point sequentially.

Proposition 3.14. *If $A \subset X$, where X is a first countable space, then x is in cls A if and only if some sequence of points in A converges to x.*

Proof. If $\{x_n\}$ is a sequence of points in A converging to x, then any open set V containing x meets the sequence and we see either $x \in \operatorname{int} A$ or $x \in \operatorname{bdy} A$, so $x \in \operatorname{cls} A$.

Conversely, if $x \in \operatorname{cls} A$, consider the collection $\{U_i^x \mid 1 = 1, 2, \dots\}$ given by the condition of first countability. Then $A \cap U_1^x \cap U_2^x \cap \cdots \cap U_n^x \neq \emptyset$ for all n. Choose some $x_n \in A \cap U_1^x \cap \cdots \cap U_n^x$. The sequence $\{x_n\}$ converges to x: If V is open in X and $x \in V$, then there is U_j^x with $x \in U_j^x \subset V$. But then $A \cap U_1^x \cap \cdots \cap U_m^x \subset U_j^x \subset V$ for all $m \geq j$, and so $x_m \in V$ for $m \geq j$. $\square$

Corollary 3.15. *In a first countable space X, a subset $A \subset X$ is closed if and only if each point of X for which $x = \lim_{n \to \infty} a_n$ for a sequence of points $a_n \in A$ satisfies $x \in A$.*

These ideas allow us to generalize the notion of sequential convergence as a criterion for continuity of functions as we will see below. In analysis it is useful to have various formulations of continuity, and so too in topology.

Theorem 3.16. *Let X, Y be topological spaces and $f \colon X \to Y$ a function. Then the following are equivalent:*

(1) *f is continuous.*

(2) *If K is closed in Y, then $f^{-1}(K)$ is closed in X.*

(3) *If $A \subset X$, then $f(\operatorname{cls} A) \subset \operatorname{cls} f(A)$.*

Proof. We first note that for any subset S of Y,

$$f^{-1}(Y - S) = \{x \in X \mid f(x) \in Y - S\}$$
$$= \{x \in X \mid f(x) \notin S\} = \{x \in X \mid x \notin f^{-1}(S)\}$$
$$= X - f^{-1}(S).$$

(1) $\Longleftrightarrow$ (2): If K is closed in Y, then $Y - K$ is open and, because f is continuous, we have $f^{-1}(Y - K) = X - f^{-1}(K)$ is open in X. Thus $f^{-1}(K)$ is closed.

If V is open in Y, then $f^{-1}(V) = X - f^{-1}(Y - V)$ and $Y - V$ is closed. So $f^{-1}(V)$ is open in X and f is continuous.

(2) $\Longrightarrow$ (3): For $A \subset X$, $\text{cls } f(A)$ is closed in Y and so $f^{-1}(\text{cls}(f(A)))$ is closed in X. It follows from $A \subset f^{-1}(f(A)) \subset f^{-1}(\text{cls } f(A))$, when $f^{-1}(\text{cls } f(A))$ is closed, that

$$\text{cls } A \subset f^{-1}(\text{cls } f(A))$$

and so $f(\text{cls } A) \subset \text{cls } f(A)$.

(3) $\Longrightarrow$ (2): If K is closed in Y, then $K = \text{cls } K$. Let $L = f^{-1}(K)$. We show $\text{cls } L \subset L$.

$$f(\text{cls } L) = f(\text{cls } f^{-1}(K)) \subset \text{cls } f(f^{-1}(K)) = \text{cls } K = K.$$

Taking inverse images, $\text{cls } L \subset f^{-1}(f(\text{cls } L)) \subset f^{-1}(K) = L$. $\qquad\square$

Part (3) of the theorem says that continuous functions send limit points to limit points.

Corollary 3.17. *If $f \colon X \to Y$ is a continuous function and $\{x_n\}$ is a sequence in X converging to x, then the sequence $\{f(x_n)\}$ converges to $f(x)$. Furthermore, if X is first countable, then the converse holds.*

Proof. Suppose $\{x_n\}$ is a sequence of points in X with $\lim_{n \to \infty} x_n = x \in X$. If $U \subset Y$ is open and $f(x) \in U$, then $x \in f^{-1}(U)$, which is open in X since f is continuous. Because $\lim_{n \to \infty} x_n = x$, there is an index N_U with $x_m \in f^{-1}(U)$ for all $m \geq N_U$. This implies that $f(x_m) \in U$ for all $m \geq N_U$ and so $\lim_{n \to \infty} f(x_n) = f(x)$.

To prove the converse, we assume that $f \colon X \to Y$ is not continuous. Then there is a closed subset of Y, $K \subset Y$, for which $f^{-1}(K)$ is not closed in X. Since the empty set is closed, we know that

$f^{-1}(K)$ and also K are not empty. Furthermore, since $f^{-1}(K)$ is not closed, there is a point $x \in \text{cls} f^{-1}(K)$ for which $x \notin f^{-1}(K)$. Because X is first countable, there is a sequence of points $\{x_n\}$ with $x_n \in f^{-1}(K)$ for all n and $\lim_{n\to\infty} x_n = x$. Then $f(x_n) \in K$ for all n and since K is closed, $\lim_{n\to\infty} f(x_n) \in K$ if it exists. However, $\lim_{n\to\infty} f(x_n) \neq f(x)$ since $x \notin f^{-1}(K)$.   □

With our general formulation of continuity, we can get a sense of the extent to which the problem of dimension is disconcerting by the following example of a continuous function due to GUISEPPE PEANO (1858–1932).

Given a real number r with $0 \leq r \leq 1$, we can represent it by its ternary expansion, that is,

$$r = 0.t_1 t_2 t_3 \cdots = \sum_{i=1}^{\infty} t_i/3^i, \text{ where } t_i \in \{0, 1, 2\}.$$

Such a representation is unique except in the special cases:

$$r = 0.t_1 t_2 \cdots t_n 222 \cdots = 0.t_1 t_2 \cdots t_{n-1}(t_n + 1)000 \cdots, \text{ where } t_n \neq 2.$$

In an 1890 paper [65], Peano introduced a function defined on $[0, 1]$ using the ternary expansion. Let σ denote the permutation of $\{0, 1, 2\}$ which exchanges 0 and 2 and leaves 1 fixed. We can think of σ as acting on the ternary digits of a number. The way in which this permutation acts can be understood by observing that when we write $r = 0.t_1 t_2 t_3 \cdots$, in its ternary expansion, then

$$1 - r = 0.222 \cdots - 0.t_1 t_2 t_3 \cdots = 0.(\sigma t_1)(\sigma t_2)(\sigma t_3) \cdots.$$

Let $\sigma^t = \sigma \circ \sigma \circ \cdots \circ \sigma$ (t times). We define

$$\text{PE}(r) = (0.a_1 a_2 a_3 \cdots, 0.b_1 b_2 b_3 \cdots)$$

by:

$$a_1 = t_1 \qquad\qquad\qquad b_1 = \sigma^{t_1} t_2$$

$$a_2 = \sigma^{t_2} t_3 \qquad\qquad\qquad b_2 = \sigma^{t_1 + t_3} t_4$$

$$\vdots \qquad\qquad\qquad\qquad \vdots$$

$$a_n = \sigma^{t_2 + t_4 + \cdots + t_{2(n-1)}} t_{2n-1} \qquad b_n = \sigma^{t_1 + t_3 + \cdots + t_{2n-1}} t_{2n}$$

$$\vdots \qquad\qquad\qquad\qquad \vdots$$

From the definition of σ and PE, the value of $\text{PE}(r)$ is the ternary expansions of a pair of real numbers $0 \leq x, y \leq 1$. The properties of the function PE prompted Hausdorff to write [33] of it: "This is one of the most remarkable facts of set theory."

Theorem 3.18. *The function* $\text{PE}: [0,1] \to [0,1] \times [0,1]$ *is well defined, continuous, and onto.*

Because this function is onto a square in $\mathbb{R}^2$, it is called a **space-filling curve**. By changing the definition of the curve slightly, it can be made to be onto, $[0,1]^{\times n} = [0,1] \times [0,1] \times \cdots \times [0,1]$ (n times) for $n \geq 2$. We note that the function is not one-one and so fails to be a bijection. However, the fact that it is continuous indicates the subtlety of the problem of dimension.

Proof. We first put the Peano curve into a form that is convenient for our discussion. The definition given by Peano is recursive and so we use this feature to give another expression for the function:

$$\text{PE}(0.t_1 t_2 t_3 \cdots) = (0.t_1, \sigma^{t_1} t_2) + (\sigma^{t_2}, \sigma^{t_1}) \circ \frac{\text{PE}(0.t_3 t_4 t_5 \cdots)}{3}.$$

Here, by $(\sigma^{t_2}, \sigma^{t_1})$, we mean the operation defined

$$(\sigma^{t_2}, \sigma^{t_1})(0.a_1 a_2 a_3 \cdots, 0.b_1 b_2 b_3 \cdots)$$
$$= (0.(\sigma^{t_2} a_1)(\sigma^{t_2} a_2)(\sigma^{t_2} a_3) \cdots, 0.(\sigma^{t_1} b_1)(\sigma^{t_1} b_2)(\sigma^{t_1} b_3) \cdots).$$

We can now prove PE is well defined. Using the recursive definition, we reduce the question of well-definedness to comparing the values $\text{PE}(0.0222 \cdots)$ and $\text{PE}(0.1000 \cdots)$ and the values $\text{PE}(0.1222 \cdots)$ and $\text{PE}(0.2000 \cdots)$. Applying the definition we find

$$\text{PE}(0.0222 \cdots) = (0.0222 \cdots, 0.222 \cdots)$$

and

$$\text{PE}(0.1000 \cdots) = (0.1000 \cdots, 0.222 \cdots).$$

The ambiguity in ternary expansions implies $\text{PE}(0.0222 \cdots) = \text{PE}(0.1000 \cdots)$.

Similarly we have

$$\text{PE}(0.1222 \cdots) = (0.1222 \cdots, 0.000 \cdots)$$

and
$$\text{PE}(0.2000\cdots) = (0.2000\cdots, 0.000\cdots),$$
and so $\text{PE}(0.1222\cdots) = \text{PE}(0.2000\cdots)$.

We next prove that the mapping PE is onto. Suppose $(u, v) \in [0, 1] \times [0, 1]$. We write
$$(u, v) = (0.a_1 a_2 a_3 \cdots, 0.b_1 b_2 b_3 \cdots).$$
Let $t_1 = a_1$. Then $t_2 = \sigma^{t_1} b_1$. Since $\sigma \circ \sigma = \text{id}$, we have $\sigma^{t_1} t_2 = \sigma^{t_1} \circ \sigma^{t_1} b_1 = b_1$. Next let $t_3 = \sigma^{t_2} a_2$. Continue in this manner to define
$$t_{2n-1} = \sigma^{t_2 + t_4 + \cdots + t_{2(n-1)}} a_n, \qquad t_{2n} = \sigma^{t_1 + t_3 + \cdots + t_{2n-1}} b_n.$$
Then $\text{PE}(0.t_1 t_2 t_3 \cdots) = (0.a_1 a_2 a_3 \cdots, 0.b_1 b_2 b_3 \cdots) = (u, v)$ and PE is onto.

Finally, we prove that PE is continuous. We use the fact that $[0, 1]$ is a first countable space and show that for all $r \in [0, 1]$, whenever $\{r_n\}$ is a sequence of points in $[0, 1]$ with $\lim_{n \to \infty} r_n = r$, then $\lim_{n \to \infty} \text{PE}(r_n) = \text{PE}(r)$.

Suppose $r = 0.t_1 t_2 t_3 \cdots$ has a unique ternary representation. For any $\epsilon > 0$, we can choose $N > 0$ with $\epsilon > 1/3^N > 0$. Then the value of $\text{PE}(r)$ is determined up to the first N ternary digits in each coordinate by the first $2N$ digits of the ternary expansion of r. For any sequence $\{r_n\}$ converging to r, there is an index $M = M(2N)$ with the property that for $m > M$, the first $2N$ ternary digits of r_m agree with those of r. It follows that the first N ternary digits of each coordinate of $\text{PE}(r_m)$ agree with those of $\text{PE}(r)$ and so $\lim_{n \to \infty} \text{PE}(r_n) = \text{PE}(r)$.

In the case that r has two ternary representations,
$$r = 0.t_1 t_2 t_3 \cdots t_N 000 \cdots = 0.t_1 t_2 t_3 \cdots (t_N - 1)222 \cdots,$$
with $t_N \neq 0$, we can apply the familiar trick (suggested by Ben Lotto) of the calculus of considering convergence from above or below the value r. Suppose that $\{r_n\}$ is a sequence in $[0, 1]$ with $\lim_{n \to \infty} r_n = r$ and $r \leq r_n$ for all n. Then for some index M, when $m > M$ we have $r_m = 0.t_1 t_2 t_3 \cdots t_N t'_{N+1} t'_{N+2} \cdots$. We can now argue as above that $\lim_{n \to \infty} \text{PE}(r_n) = \text{PE}(r)$. On the other side, for a sequence $\{s_n\}$ with $\lim_{n \to \infty} s_n = r$ and $s_n \leq r$ for all n, we compare s_n with $r = 0.t_1 t_2 t_3 \cdots (t_N - 1)222 \cdots$. Once again, we eventually have that

$s_m = 0.t_1 t_2 t_3 \cdots (t_N - 1) t''_{N+1} t''_{N+2} \cdots$. Convergence of the series $\{s_n\}$ implies that more of the ternary expansion agrees with r as n grows larger, and so $\lim_{n \to \infty} \mathrm{PE}(s_n) = \mathrm{PE}(r)$. Since convergence from each side implies general convergence, we have proved that PE is continuous. □

To get a useful picture of the Peano mapping consider the recursive expression

$$\mathrm{PE}(0.t_1 t_2 t_3 \cdots) = (0.t_1, \sigma^{t_1} t_2) + (\sigma^{t_2}, \sigma^{t_1}) \circ \frac{\mathrm{PE}(0.t_3 t_4 t_5 \cdots)}{3}.$$

When r is in the first ninth of the unit interval, we can write $r = 0.00t_3 t_4 \cdots$ and so $\mathrm{PE}(r) = \mathrm{PE}(0.t_3 t_4 t_5 \cdots)/3$. Since $0.t_3 t_4 \cdots$ varies over the entire line segment $[0, 1]$, there is a copy of the image of the interval, shrunk to fit into the lower left corner of the 3×3 subdivided square, ending at the point $(1/3, 1/3)$. The second ninth of $[0, 1]$ consists of r with $r = 0.01t_3 t_4 \cdots$ and so we find $\mathrm{PE}(r) = (0, 0.1) + (\sigma, \mathrm{id}) \circ (\mathrm{PE}(0.t_3 t_4 t_5 \cdots)/3)$. Thus the copy of the image of the interval is shrunk by a factor of 3, flipped by the mapping $(x, y) \mapsto (1 - x, y)$, a reflection across the vertical midline of the square, and then translated up by adding $(0, 0.1)$. This places the image of the origin at the point $(0.1, 0.1)$ and ties the end of the image of the first ninth to the beginning of the image of the second ninth. The well-definedness of PE is at work here.

02	10	22
01	11	21
00	12	20

If we put the first two digits of the ternary expansion of r into the appropriate subsquare, we get the pattern above and the image of the interval, shrunk to fit each subsquare, fills each subsquare oriented by the action of σ, where

$$(\sigma, \mathrm{id}) \leftrightarrow (1 - x, y); (\mathrm{id}, \sigma) \leftrightarrow (x, 1 - y); \text{ and } (\sigma, \sigma) \leftrightarrow (1 - x, 1 - y).$$

For example, the center subsquare, labeled 11, has a copy of the shrunken image of the interval upside down.

There are many approaches to space-filling curves. We have followed [65] in this exposition. Later, we will see that the failure of the Peano curve to be both onto and one-one is a feature of the topology of the unit interval and the unit square. For further discussion of the remarkable phenomenon of space-filling curves, see the book [71].

Exercises

1. Some statements about the closure operation: (1) Suppose that A is dense in X and U is open in X. Show that $U \subset \text{cls}(A \cap U)$. (2) If A, B and A_α are subsets of a topological space X, show that $\text{cls}(A \cup B) = \text{cls}(A) \cup \text{cls}(B)$. However, show that $\bigcup_\alpha \text{cls}(A_\alpha) \subset \text{cls}(\bigcup_\alpha A_\alpha)$. Give an example where the inclusion is proper. (3) Show that $\text{bdy}(A) = \text{cls}(A) \cap \text{cls}(X - A)$.

2. A subset $A \subset X$, where X is a topological space, is called **perfect** if $A = A'$, that is, A is identical with its derived set. Show that the Cantor set obtained by removing middle thirds from $[0,1]$ is a perfect subset of $\mathbb{R}$.

3. Define what it would mean for a function between topological spaces to be *continuous at a point* x in the domain.

4. A topological space X is called a **metrizable space** if the topology on X can be induced by a metric space structure on X. Not every topology on a set comes about in this fashion. Show that a metric space is always Hausdorff and first countable.

5. Suppose that X is an uncountable set and that x_0 is a given point in X. Let $\mathcal{T}_F$ denote the Fort topology on X, $\{U \mid X - U \text{ is finite or } x_0 \notin U\}$.
    i) Show that $(X, \mathcal{T}_F)$ is a Hausdorff space.
    ii) Show that $(X, \mathcal{T}_F)$ is not first countable (and hence not metrizable).

6. Suppose that (X, d) is a metric space and $A \subset X$. Define the *distance from A to a point* x, $d(x, A)$, to be the infimum of the set of real numbers $\{d(x, a) \mid a \in A\}$.
    i) Show that $d(-, A) \colon X \to \mathbb{R}$ is a continuous function.

ii) Show that a point $x \in X$ is in the closure of A if and only if $d(x, A) = 0$.

iii) What is the preimage of the closed subset $\{0\}$ of $\mathbb{R}$ under the mapping $d(-, A)$?

7. Prove that the following are topological properties: (1) X is a separable space. (2) X satisfies the Hausdorff condition. (3) X has the discrete topology.

8. An interesting problem set by Kuratowski in 1922 is called the closure-complement problem. Let X be a topological space and A a subset of X. We can apply the operations of closure, $A \mapsto \operatorname{cls} A$, and complement, $A \mapsto X - A$. By composing these operations we may obtain new subsets of X, such as the $X - \operatorname{cls} A$. Show that there are only 14 distinct such composites and that there is a subset of $\mathbb{R}^2$ for which all 14 composites are in fact distinct.

9. Show that the Peano mapping PE is not injective.

Chapter 4

Building New Spaces
from Old

The use of figures is, above all, then, for the pur-
pose of making known certain relations between the
objects that we study, and these relations are those
which occupy the branch of geometry that we have
called Analysis Situs, . . .

HENRI POINCARÉ, 1895

Having introduced topologies on sets and continuous functions, it would be useful to know what new topological spaces can be formed from a given space or spaces using certain set-theoretic constructions. The principal examples are:

1) the formation of subsets,

2) the formation of products, and

3) the formation of quotients by equivalence relations.

In later chapters, we will also introduce function spaces. In all cases we will be guided by the need for naturally occurring functions to be continuous.

Subspaces

Many interesting mathematical objects are subsets of Euclidean space, which is a topological space—how are these subsets topological spaces? By restricting the metric to a subset, it becomes a metric space and so has a topology. However, this procedure does not generalize to all topological spaces. We need a more flexible approach.

For any subset A of a set X, we associate the function $i: A \to X$ given by $i(a) = a$ (*the inclusion*). Restriction of a function $f: X \to Y$ to the subset A becomes a composite $f|_A = f \circ i: A \to Y$. To topologize a subset A of X, a topological space, we want that restriction to A of a continuous function on X be continuous. This is accomplished by giving A a topology for which $i: A \to X$ is continuous.

Definition 4.1. Let X be a topological space with topology $\mathcal{T}$ and A, a subset of X. The **subspace topology** on A is given by $\mathcal{T}_A = \{U \cap A \mid U \in \mathcal{T}\}$, also called the **relative topology** on A.

Proposition 4.2. *The collection $\mathcal{T}_A$ is a topology on A and with this topology the inclusion $i: A \to X$ is continuous.*

Proof. If U is open in X, then $i^{-1}(U) = U \cap A$, which is open in A. The fact that $\mathcal{T}_A$ is a topology on A is easy to prove and, in fact, it is the smallest topology on A making $i: A \to X$ continuous. We leave it to the reader to prove these assertions. $\qquad\square$

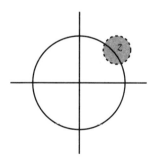

Example 1. Some interesting spaces are the **spheres** in $\mathbb{R}^n$, for $n \geq 1$. They are given by

$$S^{n-1} = \{\mathbf{x} \in \mathbb{R}^n \mid \|\mathbf{x}\| = 1\}.$$

Thus $S^0 = \{-1, 1\} \subset \mathbb{R}$, and $S^1 \subset \mathbb{R}^2$ is the unit circle. Open sets in S^1 are easy to picture: the intersection of an open ball in $\mathbb{R}^2$ with S^1 gives a sort of 'interval' in S^1. To be precise, take any point $z \in S^1$ with $z = (\cos\theta_0, \sin\theta_0)$, and let $w\colon (-\epsilon, \epsilon) \to S^1$ be the mapping $r \mapsto (\cos(\theta_0 + r), \sin(\theta_0 + r))$. Then let $\rho = d(z, (\cos(\theta_0 + \epsilon), \sin(\theta_0 + \epsilon)))$. For small ϵ, we get $w^{-1}(B(z, \rho)) = (-\epsilon, \epsilon)$ and the mapping w is a homeomorphism. Thus each point of S^1 has a neighborhood around it homeomorphic to an open set in $\mathbb{R}$. This condition is special and characterizes S^1 as a 1-dimensional manifold. More on this later.

Example 2. Some interesting subspaces of $\mathbb{R}^3$ are pictured here: they are the cylinder and the Möbius band. (Are they homeomorphic?)

 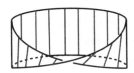

If a space X has a topological property, does a subset A of X as a subspace share it? Such a property is called **hereditary**.

Proposition 4.3. *Metrizability is a hereditary property. The Hausdorff condition is also hereditary.*

Proof. That metrizability is hereditary is left to the reader to prove. To see how the Hausdorff condition is hereditary, suppose $a, b \in A$. Then a, b are also in X, which is Hausdorff. So there are open sets U, V in X with $a \in U$, $b \in V$, and $U \cap V = \emptyset$. Consider $U \cap A$ and $V \cap A$. Since these are nonempty, disjoint, open sets in A with $a \in U \cap A$ and $b \in V \cap A$, we have that A is Hausdorff. $\qquad\square$

Reversing the notion of a hereditary property, we consider properties that, when they hold on a subspace, can be seen to hold on the whole space. For example, one can build continuous mappings this way:

Theorem 4.4. *Suppose $X = A \cup B$ is a space, A, B are open subsets of X, and $f\colon A \to Y$, $g\colon B \to Y$ are continuous functions (where A and B have the subspace topologies). If $f(x) = g(x)$ for all $x \in A \cap B$,* *then $F = f \cup g \colon X \to Y$ is a continuous function, where F is defined by*

$$F(x) = \begin{cases} f(x), & \text{if } x \in A, \\ g(x), & \text{if } x \in B. \end{cases}$$

Proof. The condition that f and g agree on $A \cap B$ implies that F is well defined. Let U be open in Y and consider

$$F^{-1}(U) = (f^{-1}(U) \cap A) \cup (g^{-1}(U) \cap B).$$

The subset $f^{-1}(U) \cap A$ is open in A so it equals $V \cap A$, where V is open in X. But since A is open, $V \cap A$ is open in X, so $f^{-1}(U) \cap A$ is open in X. Similarly $g^{-1}(U) \cap B$ is open in X and their union is $F^{-1}(U)$. Thus F is continuous. $\square$

If a space breaks up into disjoint open pieces, then continuity of a function defined on the whole space is determined by continuity on each piece.

There is a similar characterization for A, B closed in X. A subset $K \subset A$ is closed in A if there is an $L \subset X$ closed in X with $K = L \cap A$. To see this write $A - K = A \cap (X - L)$.

More generally, when A is a subspace of X and $f \colon A \to Y$ is a continuous function, is there an extension of f to all of X, $\hat{f} \colon X \to Y$, that is continuous, for which $f = \hat{f} \circ i$? This problem is called the **extension problem** and it is a common formulation of many problems in topology. An example where it is known to fail is the inclusion

$$i \colon S^{n-1} \to e^n = \operatorname{cls} B(\mathbf{0}, 1) = \{ \mathbf{x} \in \mathbb{R}^n \mid \|\mathbf{x}\| \leq 1 \} \subset \mathbb{R}^n,$$

with respect to the mapping $\operatorname{id} \colon S^{n-1} \to S^{n-1}$ (Brouwer Fixed Point Theorem in Chapter 11). The corollaries of this failure are numerous.

An extension problem with a positive solution is the following result.

Tietze Extension Theorem. *Any continuous function $f \colon A \to \mathbb{R}$ from a closed subspace A of a metric space (X, d) has an extension $g \colon X \to \mathbb{R}$ that is also continuous.*

We first prove a couple of lemmas.

Lemma 4.5. *For A a closed subset of (X, d), a metric space, let $d(x, A) = \inf\{d(x, a) \mid a \in A\}$. Then the function $x \mapsto d(x, A)$ is continuous on X.*

This is left to the reader to prove.

Lemma 4.6. *If A and B are disjoint closed subsets of (X, d), there is a real-valued continuous function in X with value 1 on A, -1 on B and values in $(-1, 1) \subset \mathbb{R}$ on $X - (A \cup B)$.*

Proof. Consider the function

$$g(x) = \frac{d(x, B) - d(x, A)}{d(x, A) + d(x, B)}.$$

Because A and B are disjoint and closed, $d(x, A) + d(x, B) > 0$ and $g(x)$ is well defined. By Lemma 4.5 and the usual theorems of real analysis, $g(x)$ is continuous, and it is rigged to satisfy the statement of the lemma. $\square$

Proof of Tietze's Theorem ([**61**, p. 212]). We first suppose $|f(x)| \leq M$ for all $x \in A$. Define

$$A_1 = \{x \in A \mid f(x) \geq M/3\}, \quad B_1 = \{x \in A \mid f(x) \leq -M/3\};$$

A_1 and B_1 are closed in A and hence in X. By Lemma 4.6, there is a continuous mapping $g_1 \colon X \to [-M/3, M/3]$ with $g_1(a) = M/3$ for $a \in A_1$, $g_1(b) = -M/3$ for $b \in B_1$, and taking values in $(-M/3, M/3)$ on $X - (A_1 \cup B_1)$. Since $|f(x)| \leq M$, $|f(x) - g_1(x)| \leq 2M/3$ for $x \in A$.

Next consider $f(x) - g_1(x)$ on A and define

$$A_2 = \{x \in A \mid f(x) - g_1(x) \geq 2M/9\},$$
$$B_2 = \{x \in A \mid f(x) - g_1(x) \leq -2M/9\}.$$

As above A_2, B_2 are closed and disjoint and so there is a continuous function $g_2 \colon X \to [-2M/9, 2M/9]$ with $g_2(a) = 2M/9$ for $a \in A_2$, $g_2(b) = -2M/9$ for $b \in B_2$, and taking values in $(-2M/9, 2M/9)$ on $x \in X - (A_2 \cup B_2)$. Notice, for $x \in A$, $|f(x) - g_1(x) - g_2(x)| \leq 4M/9$.

Iterate this process to get $g_n \colon X \to [-2^{n-1}M/3^n, 2^{n-1}M/3^n]$ such that

i) $|f(x) - g_1(x) - g_2(x) - \cdots - g_n(x)| \leq 2^n M/3^n$ on A,

ii) $|g_n(x)| < 2^{n-1}M/3^n$ on $X - A$.

For all $x \in X - A$, the infinite series satisfies

$$\left|\sum_{n=1}^{\infty} g_n(x)\right| \leq \sum_{n=1}^{\infty} |g_n(x)| \leq M\sum_{n=1}^{\infty} 2^{n-1}/3^n = M,$$

and so $g(x) = \sum_{n=1}^{\infty} g_n(x)$ converges absolutely and hence converges, defining g on $X - A$. Furthermore, $g(x) = f(x)$ for $x \in A$, and so $g(x)$ is defined for all $x \in X$; also, $|g(x)| < M$ on X and g is bounded.

To show that g is continuous, let $x_0 \in X$. We show that for any $\epsilon > 0$ there is a $\delta > 0$ such that whenever $d(x_0, x) < \delta$, then $|g(x_0) - g(x)| < \epsilon$. Define $s_n(x) = \sum_{k=1}^{n} g_k(x)$, the nth partial sum of $g(x)$. Since, for all $x \in X - A$,

$$|g(x) - s_n(x)| = \left|\sum_{k=n+1}^{\infty} g_k(x)\right| \leq \sum_{k=n+1}^{\infty} |g_k(x)|$$
$$\leq \sum_{k=n+1}^{\infty} 2^{k-1}M/3^k = M(2/3)^n,$$

then there is an N for which $|g(x) - s_n(x)| < \epsilon/3$ for $n \geq N$. On A, $|g(a) - s_n(a)| = |f(a) - s_n(a)| < 2^n M/3^n$, and so there is an N' with $|f(a) - s_n(a)| < \epsilon/3$ for $n \geq N'$. Let $N_1 = \max\{N, N'\}$.

Since $s_n(x)$ is a finite sum of continuous functions, for each n there is a $\delta_n > 0$ for which $|s_n(x_0) - s_n(y)| < \epsilon/3$ whenever $d(x_0, y) < \delta_n$. Suppose that $L > N_1$. Then, for all $y \in X$ with $d(x_0, y) < \delta_L$, we have

$$|g(x_0) - g(y)| = |g(x_0) - s_L(x_0) + s_L(x_0) - s_L(y) + s_L(y) - g(y)|$$
$$\leq |g(x_0) - s_L(x_0)| + |s_L(x_0) - s_L(y)| + |g(y) - s_L(y)|$$
$$< \epsilon.$$

Thus, for any $x_0 \in X$, g is continuous at x_0, and so g is continuous.

For an unbounded mapping $f\colon A \to \mathbb{R}$, apply the invertible mapping $h\colon \mathbb{R} \to (-1, 1)$ given by $h(r) = (2/\pi)\arctan(r)$. Let $F = h \circ f$. Then F is bounded and we can carry out the argument for F as in the bounded case to get G on X, with codomain $(-1, 1)$. Let $g = h^{-1} \circ G$. On A,

$$g = h^{-1} \circ G = h^{-1} \circ F = h^{-1} \circ h \circ f = f,$$

so g extends f to all of X. $\square$

The manner in which a subspace sits inside a larger space determines new things about the space. For example, one can make a circle a subspace of $\mathbb{R}^3$ in many ways:

The study of such embeddings is another important part of topology called *knot theory* (see [**1**], [**12**]).

One way to focus on a subspace within a space is through the continuous functions.

Definition 4.7. A **topological pair** is a space X together with a subspace A, written (X, A). A **mapping of pairs** (a continuous function of pairs) $f\colon (X, A) \to (Y, B)$ is a continuous function $f\colon X \to Y$ satisfying the additional property $f(A) \subset B$.

A composite of mappings of pairs gives a mapping of pairs and the identity mapping on a pair is a mapping of pairs. Two pairs are *homeomorphic* if there is a mapping of pairs $f\colon (X, A) \to (Y, B)$ with $f\colon X \to Y$ a homeomorphism and $f|_A\colon A \to B$ another homeomorphism. The notion of equivalence of knots reduces to whether there is a homeomorphism of pairs $(\mathbb{R}^3, K) \to (\mathbb{R}^3, K')$, where K and K' are knots, the images of homeomorphisms of S^1 with subspaces of $\mathbb{R}^3$.

A particular example of a topological pair is a pointed space.

Definition 4.8. Given a space X, a **basepoint** for X is a choice of point x_0 in X. We denote the pair $(X, \{x_0\}) = (X, x_0)$, and call (X, x_0) a **pointed space**. The mappings $f\colon (X, x_0) \to (Y, y_0)$ of such pairs are called **pointed maps**.

Example. Let $[0, 1] \subset \mathbb{R}$ with the usual topology denote the *unit interval*. A **path** in a space X is a continuous function $f\colon [0, 1] \to X$.

Choose $0 \in [0,1]$ as basepoint and define the set

$$PX = \mathrm{Hom}(([0,1],0),(X,x_0))$$
$$= \{f \colon [0,1] \to X \mid f(0) = x_0, f \text{ continuous}\},$$

the *set of all paths in X beginning at x_0*. We can also consider the set of mappings of pairs $\Omega(X,x_0) = \mathrm{Hom}(([0,1],\{0,1\}),(X,x_0))$, the *set of all paths in X beginning and ending at x_0*, also called the **loops** on X based at x_0. The loops could be described equally well as $\mathrm{Hom}((S^1,1),(X,x_0))$, where S^1 is the circle in $\mathbb{R}^2 = \mathbb{C}$ and $1 = e^{i \cdot 0} = 1 + 0i$ is chosen as basepoint for S^1. More on this set in Chapter 7.

Products

Take a pair of topological spaces X, Y and form their cartesian product:

$$X \times Y = \{(x,y) \mid x \in X, y \in Y\}.$$

How can this set be topologized to get a new space? Such a topology should make the associated projection functions continuous, namely,

$$\mathrm{pr}_1 \colon X \times Y \to X, \qquad \mathrm{pr}_2 \colon X \times Y \to Y.$$

If U is open in X, then $\mathrm{pr}_1^{-1}(U) = U \times Y$. Similarly, if V is open in Y, then $\mathrm{pr}_2^{-1}(V) = X \times V$. At the very least, we need the collection

$$\mathcal{S} = \{U \times Y, X \times V \mid U \text{ open in } X, V \text{ open in } Y\}$$

to lie in our topology on $X \times Y$. In the exercises to Chapter 2, we identified collections like $\mathcal{S}$ called *subbases* for which the collection

$$\mathcal{B} = \{S_1 \cap \cdots \cap S_n \mid n \geq 1, S_i \in \mathcal{S}\}$$

forms a basis for a topology on $X \times Y$.

Definition 4.9. The **product topology** on $X \times Y$ is the topology generated by the basis $\mathcal{B} = \{U \times V \mid U \text{ open in } X, V \text{ open in } Y\}$.

To see that we have the same basis as generated by the subbasis $\mathcal{S}$ observe that $(U \times Y) \cap (X \times V) = U \times V$. Thus the projections are continuous with the product topology on $X \times Y$. More can be said:

Proposition 4.10. *Given three topological spaces X, Y, and Z, and a function $f \colon Z \to X \times Y$, then f is continuous if and only if $\mathrm{pr}_1 \circ f \colon Z \to X$ and $\mathrm{pr}_2 \circ f \colon Z \to Y$ are continuous.*

Proof. Certainly f being continuous implies $\mathrm{pr}_1 \circ f$ and $\mathrm{pr}_2 \circ f$ are continuous. To prove the converse, suppose W is an open set in $X \times Y$. Then W is a union of $U_i \times V_i$ with each U_i open in X, V_i open in Y. Since $f^{-1}(\bigcup(U_i \times V_i)) = \bigcup f^{-1}(U_i \times V_i)$, we can restrict our attention to a basis open set. The subsets $(\mathrm{pr}_1 \circ f)^{-1}(U_i)$ and $(\mathrm{pr}_2 \circ f)^{-1}(V_i)$ are both open in Z by the hypotheses. The proof reduces to proving

$$f^{-1}(U_i \times V_i) = (\mathrm{pr}_1 \circ f)^{-1}(U_i) \cap (\mathrm{pr}_2 \circ f)^{-1}(V_i) :$$

If z is in $f^{-1}(U_i \times V_i)$, then $f(z) \in U_i \times V_i$ and $\mathrm{pr}_1 \circ f(z) \in U_i$, $\mathrm{pr}_2 \circ f(z) \in V_i$. Thus $f^{-1}(U_i \times V_i) \subset (\mathrm{pr}_1 \circ f)^{-1}(U_i) \cap (\mathrm{pr}_2 \circ f)^{-1}(V_i)$. If $z \in (\mathrm{pr}_1 \circ f)^{-1}(U_i) \cap (\mathrm{pr}_2 \circ f)^{-1}(V_i)$, then $f(z) \in \mathrm{pr}_1^{-1}(U_i) \cap \mathrm{pr}_2^{-1}(V_i)$ $= U_i \times V_i$. $\qquad\square$

By induction, we can endow a finite product $X_1 \times X_2 \times \cdots \times X_n$ with a topology for which the projections $\mathrm{pr}_i \colon X_1 \times X_2 \times \cdots \times X_n \to X_i$, $\mathrm{pr}_i(x_1, \ldots, x_n) = x_i$, are continuous. Proposition 4.10 generalizes for functions $f \colon Z \to X_1 \times X_2 \times \cdots \times X_n$ that are continuous if and only if all the compositions $\mathrm{pr}_i \circ f$ are continuous. This obtains the fact from classical analysis that a function $f \colon Z \to \mathbb{R}^n$ is continuous if and only if the coordinate functions expressing f are continuous.

We had hereditary properties for subspaces; are there topological properties that go over to products when they hold for each factor? We give an example:

Proposition 4.11. *If X and Y are separable spaces, so is $X \times Y$.*

Proof. Let $A \subset X$ and $B \subset Y$ be countable dense subsets. Then $A \times B \subset X \times Y$ is also countable. To see that it is dense, suppose $(x, y) \in X \times Y$ and $(x, y) \notin A \times B$, and W is an open set in $X \times Y$ with $(x, y) \in W$. Then there is a basis open set $U \times V$ with $(x, y) \in U \times V \subset W$. Since A is dense in X, there is an $a \in A$ with $a \neq x$ and $a \in U$. Similarly there is a $b \in B$, $b \in V$, and $b \neq y$. Thus $(a, b) \in W$ with $(a, b) \neq (x, y)$. Hence (x, y) is a limit point of $A \times B$, and $\mathrm{cls}(A \times B) = X \times Y$. $\qquad\square$

Many other properties act analogously, for example, the Hausdorff condition, or second countability, and others.

We can extend the notion of products to infinite products and then try to extend the product topology to them; this requires care.

Definition 4.12. Let $\{X_\alpha \mid \alpha \in J\}$ be any collection of nonempty sets. The **product** of the sets $\prod_{\alpha \in J} X_\alpha$ is the set of all functions $c \colon J \to \bigcup_{\alpha \in J} X_\alpha$ with $c(\alpha) \in X_\alpha$ for all $\alpha \in J$. For any $\beta \in J$, the **projection** $\mathrm{pr}_\beta \colon \prod_{\alpha \in J} X_\alpha \to X_\beta$ is given by evaluation of such a function c on β, $c \mapsto c(\beta)$.

This structure describes products for any collection and generalizes finite products for which the indexing set is $\{1, 2, \ldots, n\}$. Why do we need such notions? Consider $\mathbb{R}^\omega = \{(r_1, r_2, r_3, \ldots) \text{ such that } r_i \in \mathbb{R}\}$, the countable product of $\mathbb{R}$ with itself. A nice example of a subspace of $\mathbb{R}^\omega$ is an important space in analysis that generalizes $\mathbb{R}^n$:

$$l^2 = \{\text{ square summable sequences of } \mathbb{R} \}$$
$$= \{(r_1, r_2, r_3, \ldots) \mid \sum_{i=1}^{\infty} r_i^2 < \infty\}.$$

The norm $\|(r_1, r_2, r_3, \ldots)\| = \sqrt{\sum_i r_i^2}$ provides a distance function and hence a metric space structure on l^2.

What is the infinite analogue of the product topology on $X \times Y$? Two alternatives are possible: let $\prod_{\alpha \in J} X_\alpha$ be a product of spaces $\{X_\alpha \mid \alpha \in J\}$,

i) $\mathcal{T}_{\mathrm{box}} = $ the topology generated by the basis $\mathbb{B} = \{\prod_{\alpha \in J} U_\alpha \mid U_\alpha \subset X_\alpha$ for all α, each U_α open in $X_\alpha\}$.

ii) $\mathcal{T}_{\mathrm{prod}} = $ the topology generated by the basis $\mathcal{B} = \{S_1 \cap S_2 \cap \cdots \cap S_n \mid n \geq 1, S_i \in \mathcal{S}\}$, where $\mathcal{S}$ is the subbasis of subsets $S = \prod_{\alpha \in J} V_\alpha$, where for each $\beta \in J$, V_β is open in X_β and $V_\gamma = X_\gamma$ for all but finitely many $\gamma \in J$.

Definition 4.13. The topology $\mathcal{T}_{\mathrm{box}}$ is called the **box topology** on $\coprod_{\alpha \in J} X_\alpha$. The topology $\mathcal{T}_{\mathrm{prod}}$ is called the **product topology**.

In both cases it is easy to prove we have topologies. (Check this!) Furthermore, all of the projections $\mathrm{pr}_{\alpha'} \colon \prod_{\alpha \in J} X_\alpha \to X_{\alpha'}$ are continuous in both topologies. To see the difference we observe the following: A subset W of $\prod_{\alpha \in J} X_\alpha$ is open in the product topology if it is a union of subsets of the form $\prod_{\alpha \in J} V_\alpha$, where $V_\alpha = X_\alpha$ for all but finitely many $\alpha \in J$. If J is infinite and only finitely many of the X_α are indiscrete spaces, then $\mathcal{T}_{\text{box}}$ is *strictly finer* than $\mathcal{T}_{\text{prod}}$.

A decisive difference appears when we form the product of a fixed space with itself over an index set.

Proposition 4.14. *Let X be a space and for all $\alpha \in J$, let $X_\alpha = X$. Define the function*
$$\Delta \colon X \to \prod_{\alpha \in J} X_\alpha$$
by $\Delta(x)\colon \alpha \mapsto x \in X_\alpha = X$. This function is continuous when $\prod_{\alpha \in J} X_\alpha$ has the product topology.

Proof. If $\prod_{\alpha \in J} V_\alpha$ is a basic open set, then $V_\beta = X$ for all but finitely many $\beta \in J$, say $\alpha_1, \alpha_2, \ldots, \alpha_n$. Then $\Delta^{-1}(\prod_{\alpha \in J} V_\alpha) = \bigcap_{\alpha \in J} V_\alpha = V_{\alpha_1} \cap \cdots \cap V_{\alpha_n}$, which is open in X. $\qquad\square$

Compare $\Delta \colon (\mathbb{R}, \text{usual}) \to (\mathbb{R}^\omega, \mathcal{T}_{\text{box}})$. The open set
$$(-1,1) \times (-1/2, 1/2) \times (-1/3, 1/3) \times \cdots = W$$
has $\Delta^{-1}(W) = \{0\}$, which is not open. Since the composites $\text{pr}_i \circ \Delta = $ id, a desirable property of continuous functions on products fails. This example recommends the product topology over the box topology as *the* product topology.

Another nice property of the product topology is the preservation of certain properties: for example, a product of Hausdorff spaces is Hausdorff. However, an uncountable product of second countable spaces or separable spaces need not be second countable or separable.

When spaces are pointed, $(X_\alpha, x_{\alpha 0})$, we can construct some continuous functions of interest. The product $\prod_{\alpha \in J} X_\alpha$ is pointed with basepoint $(\alpha \mapsto x_{\alpha 0})_{\alpha \in J}$. Define the injections
$$i_\alpha \colon (X_\alpha, x_{\alpha 0}) \to \left(\prod_{\beta \in J} X_\beta, (\beta \mapsto x_\beta)_{\beta \in J} \right)$$
given by $x \mapsto c$, where $c \colon J \to \bigcup_{j \in J} X_j$ is defined
$$c(j) = \begin{cases} x, & \text{if } j = \alpha, \\ x_{\alpha' 0}, & \text{if } j \neq \alpha,\ j = \alpha'. \end{cases}$$
The preimage under i_α of an open set is determined only by the open set in the coordinate α so each i_α is continuous. Notice, without

the chosen basepoints, there is no obvious way to choose the other coordinates to define the inclusions i_α.

Next, notice $\mathrm{pr}_\alpha \circ i_\alpha = \mathrm{id}\colon X_\alpha \to X_\alpha$. Thus we can factor the identity through the pointed product space.

Finally, we mention an interesting subspace of $(X \times Y, (x_0, y_0))$.

Definition 4.15. The **one-point union of the pointed spaces** (X, x_0) and (Y, y_0), denoted $X \vee Y$, is given by $X \times \{y_0\} \cup \{x_0\} \times Y \subset X \times Y$.

One can think of $X \vee Y$ as the pair of axes joined at the origin (x_0, y_0). A homeomorphic image of $S^2 \vee S^1$ can be pictured as a sphere with a circle touching it at a point.

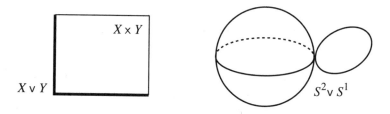

There are canonical mappings $X \to X \vee Y \to X$ given by $x \mapsto (x, y_0) \mapsto x$. When $X = Y$, the extension problem posed by taking $X \vee X \subset X \times X$ and the *fold map* fold$\colon X \vee X \to X$ given by fold$(x, x_0) = x = $ fold(x_0, x) is solved by a continuous binary operation $\mu\colon X \times X \to X$ for which x_0 is an identity element. Spaces like this are called *H-spaces* (or *Hopf spaces*). They are generalizations of groups and they play an important role in topology.

Quotients

Another method for building new spaces starts with a space X and an equivalence relation $\sim$ on X. The space X maps to the set of equivalence classes $[X]$ via the *canonical surjection* $\mathrm{pr}\colon X \to [X]$, $x \mapsto [x]$, the equivalence class of x. We want to introduce a topology on $[X]$ which makes the canonical surjection continuous. We take the most direct course.

Definition 4.16. A subset $V \subset [X]$ is open in the **quotient topology** on $[X]$ if $\mathrm{pr}^{-1}(V)$ is open in X. The space $[X]$ with this topology is called a **quotient space** of X.

Notice that the quotient topology is the *finest* topology making $\mathrm{pr}\colon X \twoheadrightarrow [X]$ continuous: anything larger would have open sets whose preimage would not be open. We characterize the relation between the quotient topology and the canonical surjection.

Definition 4.17. An onto map $f\colon X \to Y$ is called a **quotient map** when V is open in Y if and only if $f^{-1}(V)$ is open in X.

Observation. Some continuous functions $f\colon X \to Y$ enjoy a more unlikely property; $f(U) \subset Y$ is open when U is open in X. Such continuous mappings are called **open mappings**; there is also the analogous notion of a closed mapping. A homeomorphism is open as is a canonical projection from a product.

Theorem 4.18. (1) *If $f\colon X \to Y$ is an onto, continuous mapping, then f is a quotient map if it is an open mapping.* (2) *If $f\colon X \to Y$ is a quotient map, then a function $g\colon Y \to Z$ is continuous if and only if the composite $g \circ f\colon X \to Z$ is continuous.* (3) *Suppose $f\colon X \to Y$ is a quotient map. Suppose $\sim$ is the equivalence relation defined in X by $x \sim x'$ if $f(x) = f(x')$. Then the quotient space $[X]$ is homeomorphic to Y.*

Proof. (1) We need to show that f is an open mapping implies f is a quotient map. Suppose V is any subset in Y. Then, if $f^{-1}(V)$ is open in X, $f(f^{-1}(V)) = V$ is open in Y since f is an open mapping. Hence f is a quotient map.

(2) We need to show that $g \circ f$ is continuous implies g is continuous. Suppose W is open in Z. Then $(g \circ f)^{-1}(W) = f^{-1}(g^{-1}(W))$ is open in X. Since f is a quotient map, $g^{-1}(W)$ is open in Y. Hence, g is continuous.

(3) Since we have an equivalence relation, we consider the diagram

$$\begin{array}{ccc} X & \xrightarrow{f} & Y \\ \downarrow{\scriptstyle\text{pr}} & & \| \\ [X] & \xrightarrow[\hat{f}]{} & Y \end{array}$$

The lift $\hat{f}\colon [X] \to Y$ is given by $\hat{f}([x]) = f(x)$ and it is well defined by the conditions of (3). Notice that $\hat{f} \circ \text{pr} = f$. Both f and pr are quotient maps so $\hat{f}$ is continuous. We show that $\hat{f}$ is one-one and onto and $\hat{f}^{-1}$ is continuous, which implies that $\hat{f}$ is a homeomorphism. If $\hat{f}([x]) = \hat{f}([x'])$, then $f(x) = f(x')$ and so $x \sim x'$, that is, $[x] = [x']$, and $\hat{f}$ is one-one. If $y \in Y$, then $y = f(x)$ since f is onto and $\hat{f}([x]) = y$ so $\hat{f}$ is onto. To see that $\hat{f}^{-1}$ is continuous, observe that since f is a quotient map and pr is a quotient map, this shows $\text{pr} = \hat{f}^{-1} \circ f$ and (2) implies that $\hat{f}^{-1}$ is continuous.   □

Part (3) of Theorem 4.18 allows useful comparisons. Let's consider an example.

Example. Let $\sim$ be the equivalence relation on $\mathbb{R}$ given by $r \sim s$ if $s - r$ is an integer. Give $\mathbb{R}$ the usual topology and consider $[\mathbb{R}]$. Intuitively we have identified two real numbers whenever they differ by an integer and so only $[0, 1]$ would be in $[\mathbb{R}]$ with $0 \sim 1$. That is, *form the space from* $[0, 1]$ *by joining* 0 *to* 1. This ought to be a circle! Consider the mapping

$$f\colon \mathbb{R} \to S^1, \qquad f(r) = (\cos(2\pi r), \sin(2\pi r)).$$

If $r \sim s$, then $f(r) = f(s)$ so we get a function $\hat{f}\colon [\mathbb{R}] \to S^1$, such that the following diagram commutes:

$$\begin{array}{ccc} \mathbb{R} & \xrightarrow{f} & S^1 \\ \downarrow{\scriptstyle\text{pr}} & & \| \\ [\mathbb{R}] & \xrightarrow[\hat{f}]{} & S^1 \end{array}$$

From calculus we know f is continuous and $f = \hat{f} \circ \text{pr}$, so by Theorem 4.18(2) $\hat{f}$ is continuous. Furthermore $\hat{f}$ is one-one and onto, so we only need to know if $\hat{f}$ is open to see that it is a homeomorphism. We could apply (3) above more easily if f were open, so we check: let $(a, b) \subset \mathbb{R}$, $a < b$, be a basic open set. The image $f((a, b))$ consists of those points on S^1 of angle between $2\pi a$ and $2\pi b$, which is open in S^1. Thus f is open and $[\mathbb{R}] \cong S^1$.

Quotient spaces let us make precise a construction called **glue-ing**. Suppose one has two subsets $A, B \subset X$ and a homeomorphism $h : A \to B$. We can define the equivalence relation $\sim_h$ on X by $x \sim_h x'$ if $x = x'$, $h(x) = x'$, or $h^{-1}(x) = x'$. This identifies points $a \in A$ with their counterpart $h(a) \in B$ and vice versa. This process 'glues' A to B according to h. Let's consider some specific examples.

(1) Let $I^2 = [0, 1] \times [0, 1]$ and define $A = \{0\} \times [0, 1] \cup [0, 1] \times \{0\}$ and $B = \{1\} \times [0, 1] \cup [0, 1] \times \{1\}$; then take the mapping $h : A \to B$ by $h((0, t)) = (1, t)$ and $h((t, 0)) = (t, 1)$. This glues the bottom of the box to the top and the sides to the sides. We get a *torus* in this fashion given as in the diagram:

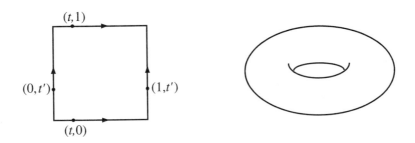

Alternatively, the torus can be described as a circle rotated around a line outside it. Taking the coordinates of a point on the torus from the given circle and the rotation shows the torus $T^2 = S^1 \times S^1$. This description leads to a function $f : I^2 \to T^2$ given by $f(u, v) = (e^{2\pi i u}, e^{2\pi i v}) \in S^1 \times S^1$. Since $e^{2\pi i 0} = e^{2\pi i 1}$ we get $f(u, v) = f(\bar{u}, \bar{v})$ if and only if $(u, v) \sim_h (\bar{u}, \bar{v})$. Thus we get $\hat{f} : [I^2]_h \to T^2$, which is a homeomorphism in the same way as in the argument for the circle.

(2) The following famous quotient of a square was constructed in 1858 independently by Johann Listing, who introduced the word 'topology' for such studies, and Möbius for whom it is named. Let

$X = [0,1] \times [0,1]$ and let $A = \{0\} \times [0,1]$, $B = \{1\} \times [0,1]$ with the homeomorphism $h(0,t) = (1, 1-t)$. Then $[X]_h$ represents the Möbius band, M. From a convenient representation of M in $\mathbb{R}^3$, the quotient map is evident.

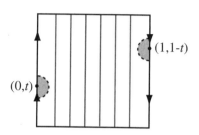 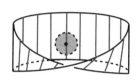

Notice how an open set around a point on the line segment where it is glued has preimage an open set (in two pieces) in X.

(3) One of the most important spaces in topology is the **projective plane**. Its formal definition is given as a set by

$$\mathbb{R}P^2 = \{\, \text{lines through the origin in } \mathbb{R}^3 \,\}.$$

To 'tame' this description a bit, we introduce coordinates for a point in $\mathbb{R}P^2$. Suppose $(x,y,z) \in \mathbb{R}^3$ and $(x,y,z) \neq (0,0,0)$. Introduce the equivalence relation $(x,y,z) \sim (\lambda x, \lambda y, \lambda z)$ for $\lambda \in \mathbb{R} - \{0\}$. Then $\mathbb{R}P^2 = [\mathbb{R}^3 - \{0\}]$, topologized as a quotient space.

The projective plane is the home for algebraic curves, defined as zero sets of homogeneous polynomials in two variables. The fact that such an algebraic curve lies in $\mathbb{R}P^2$ provides further geometry with which to study the curve. Also, projective geometry is modelled by the projective plane.

We construct a more easily described topological model for $\mathbb{R}P^2$: To each line in $\mathbb{R}^3$ through the origin, we can associate two points $\{\pm(x,y,z)\}$ in S^2 by taking the two points of intersection of the line with the sphere. The inclusion $S^2 \hookrightarrow \mathbb{R}^3 - \{0\}$ composed with the canonical surjection pr$\colon \mathbb{R}^3 - \{0\} \to [\mathbb{R}^3 - \{0\}]$ gives a mapping $S^2 \to \mathbb{R}P^2$ and we get the associated equivalence relation on S^2 as $(x,y,z) \sim (x',y',z')$ whenever $(x',y',z') = \pm(x,y,z)$. Thus $\mathbb{R}P^2 \cong [S^2]$, where we identify antipodal points together. A *projective line* is the image of the intersection of a plane through the origin with S^2 (a great circle) in $\mathbb{R}P^2$. Two points on $\mathbb{R}P^2$ determine a unique projective line by taking the plane spanned by the points and the origin in $\mathbb{R}^3$, and two projective lines meet in the line given by the intersection of the planes that determine them.

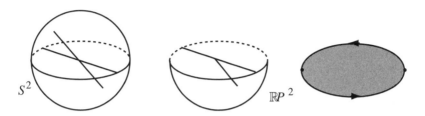

The hemisphere in the picture tells us how to represent $\mathbb{R}P^2$ as a quotient of a disk: On the rim of the hemisphere antipodal points are identified—this is the line at infinity in the projective plane. So let

$$e^2 = \{(x, y) \mid x^2 + y^2 \le 1\} \subset \mathbb{R}^2$$

be the 2-disk. Let $A = \{(x, y) \mid x^2 + y^2 = 1 \text{ and } x \ge 0\}$, $B = \{(x, y) \mid x^2 + y^2 = 1 \text{ and } x \le 0\}$ and define $h: A \to B$ by $h(x, y) = (-x, -y)$. The quotient space $[e^2]_h$ is, once again, $\mathbb{R}P^2$.

All of this discussion generalizes to define $\mathbb{R}P^n$, the n-**dimensional projective space**, as $[S^n]$ where $\mathbf{x} \sim \pm\mathbf{x}$. These spaces are the object of intense study in modern topology.

Here are some standard constructions that apply to any space X.

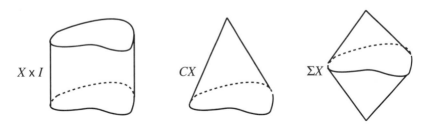

(4) The **cone** on X is given by $[X \times [0, 1]]$ where $(x, t) \sim (x', t')$ if $(x, t) = (x', t')$ or $x, x' \in X$ and $t = t' = 0$. We write $CX = [X \times [0, 1]]$ for the cone on X.

(5) The **suspension** of X, denoted ΣX, is the quotient of $X \times [0,1]$, where we identify the subsets $X \times \{0\}$ and $X \times \{1\}$ each to a point (two points here).

Suspension gives a convenient construction of the spheres:

Theorem 4.19. *The $(n+1)$-sphere S^{n+1} is homeomorphic to ΣS^n.*

Proof. Consider the function $\sigma \colon S^n \times [0,1] \to S^{n+1}$ given by

$$\sigma(x_0, \ldots, x_n, t) = (\sqrt{1-(1-2t)^2}x_0, \ldots, \sqrt{1-(1-2t)^2}x_n, 1-2t).$$

This function is continuous as the calculus tells us. Notice that

$$\sigma(x_0, \ldots, x_n, 0) = (0,0,\ldots,0,1), \quad \sigma(x_0,\ldots,x_n,1) = (0,0,\ldots,0,-1).$$

Thus σ factors through $[S^n \times [0,1]] = \Sigma S^n$.

$$\begin{array}{ccc}
S^n \times [0,1] & \xrightarrow{\ \sigma\ } & S^{n+1} \\
{\scriptstyle \mathrm{pr}}\downarrow & & \| \\
{[S^n \times [0,1]]} & \xrightarrow{\ \hat{\sigma}\ } & S^{n+1}.
\end{array}$$

The function $\hat{\sigma}$ is one-one, onto away from the 'poles' $(0,\ldots,0,\pm 1)$. The classes remaining, $[S^n \times \{0\}]$ and $[S^n \times \{1\}]$, each go to the respective poles. To finish the proof we only need to show that σ is a quotient map. Let $S^n \times [0,1]$ get its topology as a subspace of $\mathbb{R}^{n+2}$. A basic open set in $S^n \times [0,1]$ takes the form $W = (S^n \times [0,1]) \cap [(a_1,b_1) \times \cdots \times (a_{n+2}, b_{n+2})]$. Restricting (or extending) σ to W takes it to an open set and the image is easily determined to be the intersection of $\sigma(W)$ with S^{n+1}. Thus σ is open. $\qquad\square$

There are pointed versions of CX and ΣX: Given (X, x_0), a pointed space, then $(\tilde{C}X, Cx_0)$ is $[CX] = [X \times [0,1]]_\approx$, where $(x,t) \approx (x',t')$ if $(x,t) = (x',t')$, or $t = 0$, x, $x' \in X$ or $x - x' = x_0$ and $t \in [0,1]$. The single class Cx_0 in $[X \times [0,1]]_\approx$ is given by the subset $\{(x,0), x \in X, (x_0,t), t \in [0,1]\}$.

The pointed suspension (SX, sx_0) has $[sx_0] = X \times \{0\} \cup X \times \{1\} \cup x_0 \times [0,1]$, and the rest of the equivalence classes are the same as for ΣX. An extraordinary property of SX is the following.

Proposition 4.20. *There is a one-one correspondence of sets*

$$\operatorname{Hom}((SX, sx_0), (Y, y_0)) \cong \operatorname{Hom}((X, x_0), \operatorname{Hom}((S^1, 1), (Y, y_0))).$$

Proof. Let $f \colon (SX, sx_0) \to (Y, y_0)$. Untangling the suspension coordinate we can write f in the composite

$$X \times [0, 1] \xrightarrow{\operatorname{pr}} SX \xrightarrow{f} Y$$

and for each $x \in X$ associate the mapping $x \mapsto \tilde{f}(t) = f \circ \operatorname{pr}(x, t)$. It follows that $\tilde{f}(0) = \tilde{f}(1) = f(sx_0) = y_0$ by the definition of the canonical projection for the equivalence relation. The inverse is as follows: given $F \colon (X, x_0) \to \operatorname{Hom}((S^1, 1), (Y, y_0))$, then define $\hat{F} \colon (SX, sx_0) \to (Y, y_0)$ by $\tilde{F}(x, t) = F(x)(e^{2\pi i t})$. An explicit calculation shows these processes to be inverses and the proposition is proved. □

Are certain topological properties respected by quotient maps? One must be careful. For example, we can partition $(\mathbb{R}, \text{usual})$ into three parts $A = (-\infty, 0)$, $B = \{0\}$, $C = (0, \infty)$. The associated quotient is a three-point set $X = \{a, b, c\}$ for the equivalence classes and topology $\{\emptyset, X, \{a, b\}, \{a\}, \{b\}\}$, where $a = [A]$, $b = [B]$, and $c = [C]$. However, this topology is *not* Hausdorff! More can be said however.

Theorem 4.21. *Let $\sim$ be an equivalence relation in a space X that is Hausdorff. Then $[X]$ is Hausdorff if and only if the **graph of** $\sim$, $\{(x, y) \mid x \sim y, x, y \in X\}$, is closed in $X \times X$.*

Proof. Let $[x], [y] \in [X]$ and $[x] \neq [y]$. Then the point $(x, y) \in X \times X$ lies outside the graph of $\sim$, which is closed. Choose a basic open set $U \times V \subset X \times X$ with $x \in U$, $y \in V$, and $U \times V \subset X \times X - \operatorname{graph}(\sim)$. Consider $\operatorname{pr}(U) \subset [X]$. Then $[x] \in \operatorname{pr}(U)$ and similarly $[y] \in \operatorname{pr}(V)$. We claim that $\operatorname{pr}(U)$ and $\operatorname{pr}(V)$ are open and disjoint. Openness follows from the fact that pr is an open mapping. Suppose $[w] \in \operatorname{pr}(U) \cap \operatorname{pr}(V)$. Then there is a point v with $v \sim w$ and a point $v' \sim w$ with $v \in U$, $v' \in V$. But then $(v, v') \in U \times V$ and so $U \times V \cap \operatorname{graph}(\sim) \neq \emptyset$; a contradiction. This shows $[X]$ is Hausdorff. The converse is left to the reader. □

Exercises

1. Prove Lemma 4.5.

2. Show that a space X is Hausdorff if and only if the subset $\Delta(X) = \{(x,x) \mid x \in X\}$ is a closed subset of the product space $X \times X$. Suppose X and Y are Hausdorff spaces. Show that $X \times Y$ is also Hausdorff. Finish the proof of Theorem 4.21.

3. Suppose $X = A_1 \cup A_2 \cup \cdots$, where A_n is open in X for all n. If $f \colon X \to Y$ is a function such that, for each n, $f|_{A_n} \colon A_n \to Y$ is continuous with respect to the subspace topology on A_n, show that f is itself continuous. What is the analogous statement when X is a union of closed sets?

4. Suppose that we have two pointed spaces (X, x_0) and (Y, y_0). Show that the mappings $X \to X \times Y$, given by $x \mapsto (x, y_0)$, and $Y \to X \times Y$, given by $y \mapsto (x_0, y)$, are each continuous, and have continuous **sections**. (A function $f \colon U \to V$ has a section, g, if the function $g \colon V \to U$ is such that $g \circ f \colon U \to U$ is the identity mapping. This need not be a strict inverse as in the case above. Notice that f will be one-one, but not necessarily onto.)

5. Consider the subspace of $\mathbb{R}^2$ given by

$$S^1 = \{(x,y) \in \mathbb{R}^2 \mid x^2 + y^2 = 1\}.$$

This is the unit circle. The mapping

$w \colon [0, 1) \to S^1$ given by $w(r) = (\cos(2\pi r), \sin(2\pi r))$

is one-one and onto. Show that it is continuous if you give S^1 the subspace topology from $\mathbb{R}^2$, but that the inverse function is *not* continuous.

6. A **topological group** is a group that is a Hausdorff topological space, and the binary operation $\mu \colon G \times G \to G$ and the mapping $x \mapsto x^{-1}$ are continuous.

   i) Prove that a group G is a topological group if and only if it is a Hausdorff topological space and the mapping $G \times G \to G$ given by $(x, y) \mapsto x^{-1} \cdot y$ is continuous.

    ii) Let g_0 be an element of a topological group G. Show that the mappings $R_{g_0}: G \to G$ and $L_{g_0}: G \to G$ given by $R_{g_0}(h) = h \cdot g_0$ and $L_{g_0}(h) = g_0 \cdot h$ are homeomorphisms of G with itself.

    iii) Prove that the reals with addition is a topological group, and the nonzero reals with multiplication form a group. This amounts to showing that $+$ and $\times$ are continuous on $(\mathbb{R}, \text{usual})$. Do this in detail.

7. Recall that the projective plane is defined to be the set of lines in $\mathbb{R}^3$ through the origin. There is also a representation of $\mathbb{R}P^2$ as a quotient of the 2-sphere S^2 by identifying antipodal points:

    i) Let $S^2 \cong D^+ \cup C \cup D^-$, where D^+ is the part above and on the plane $z = \frac{1}{2}$, D^- is the part on and below the plane given by $z = -\frac{1}{2}$, and C is the part in between. Let $p: S^2 \to \mathbb{R}P^2$ be the quotient map. Verify that $D^+ \cong D^2 = \{\mathbf{x} \in \mathbb{R}^3 \mid \|\mathbf{x}\| \leq 1\}$. Verify that $C \cong S^1 \times [0,1]$. And verify by cutting and glueing that $p(C)$ is homeomorphic to a Möbius band embedded in $\mathbb{R}P^2$.

    ii) Verify that $p(D^+) \cup p(C) = \mathbb{R}P^2$ and that $p(C) \cap p(D^+) \cong S^1$. This shows that the projective plane can be obtained from attaching a disk to the Möbius band along its edge.

8. Suppose that $A \subset X$ is a nonempty closed subset of a space X that is Hausdorff, and further that X satisfies the property that if $x \in X$ and $x \notin A$, then there are open sets U and V with $x \in U$, $A \subset V$, and $U \cap V = \emptyset$. Define the relation $x \sim y$ if $x = y$ or x and $y \in A$. Show that this relation is an equivalence relation. The quotient topology on $[X]$ is denoted by the space X/A. Show that the quotient space X/A is Hausdorff. A space that has this separation property for every closed proper subset A is said to satisfy the T_3 axiom. Show that being T_3 is a topological property.

Chapter 5

Connectedness

We begin our introduction to topology with the study of connectedness—traditionally the only topic studied in both analytic and algebraic topology.

C. T. C. WALL, 1972

The property at the heart of certain key results in analysis is that of connectedness. The definition applies to any topological space.

Definition 5.1. A space X is **disconnected** by a separation $\{U, V\}$ if U and V are open, nonempty, and disjoint ($U \cap V = \emptyset$) subsets of X with $X = U \cup V$. If no separation of the space X exists, then X is **connected**.

Notice that $V = X - U$ is closed and likewise U is closed. A subset that is both open and closed is sometimes called *clopen*. Closure leads to an equivalent condition.

Theorem 5.2. *A space X is connected if and only if whenever $X = A \cup B$ with A, B nonempty, then $A \cap (\operatorname{cls} B) \neq \emptyset$ or $(\operatorname{cls} A) \cap B \neq \emptyset$.*

Proof. If $A \cap (\operatorname{cls} B) = \emptyset$ and $(\operatorname{cls} A) \cap B = \emptyset$, then, since $A \cup B = X$, it will follow that $\{X - \operatorname{cls} A, X - \operatorname{cls} B\}$ is a separation of X. To see this, consider $x \in (X - \operatorname{cls} A) \cap (X - \operatorname{cls} B)$; then $x \notin \operatorname{cls} A$ and $x \notin \operatorname{cls} B$. But then $x \notin \operatorname{cls} A \cup \operatorname{cls} B = X$, a contradiction. Therefore $(X - \operatorname{cls} A) \cap (X - \operatorname{cls} B) = \emptyset$. Thus we have a separation.

Conversely, if $\{U, V\}$ is a separation of X, let $A = X - V = U$ and $B = X - U = V$. Since U and V are open, A and B are closed. Then $X = U \cup V = A \cup B$. However, $A \cap \text{cls } B = A \cap B = U \cap V = \emptyset$. □

Example. The canonical connected space is the unit interval $[0, 1] \subset$ $(\mathbb{R}, \text{usual})$. To see this, suppose $\{U, V\}$ is a separation of $[0, 1]$. Suppose that $0 \in U$. Let $c = \sup\{0 \leq t \leq 1 \mid [0, t] \subset U\}$. If $c = 1$, then $V = \emptyset$, so suppose $c < 1$. Since $c \in [0, 1]$, $c \in U$ or $c \in V$. If $c \in U$, then there exists an $\epsilon > 0$ such that $(c - \epsilon, c + \epsilon) \subset U$ and there is a natural number $N > 1$ such that $c < c + (\epsilon/N) < 1$. But this contradicts c being a supremum since $c + (\epsilon/N) \in [0, 1]$. If $c \in V$, then there exists a $\delta > 0$ such that $(c - \delta, c + \delta) \subset V$. For some $N' > 1$, $c + (\delta/N') < 1$ and so $(c - (\delta/N'), c + (\delta/N'))$ does not meet U, so c could not be a supremum. Since the set $\{0 \leq t \leq 1 \mid [0, t] \subset U\}$ is nonempty and bounded, it has a supremum. It follows that $c = 1$ and so $[0, 1]$ is connected.

Is connectedness a topological property? In fact more is true:

Theorem 5.3. *If $f \colon X \to Y$ is continuous and X is connected, then $f(X)$, the image of X in Y, is connected.*

Proof. Suppose $f(X)$ has a separation. It would be of the form $\{U \cap f(X), V \cap f(X)\}$ with U and V open in Y. Consider the open sets $\{f^{-1}(U), f^{-1}(V)\}$. Since $U \cap f(X) \neq \emptyset$, we have $f^{-1}(U) \neq \emptyset$ and similarly $f^{-1}(V) \neq \emptyset$. Since $U \cap f(X) \cup V \cap f(X) = f(X)$, we have $f^{-1}(U) \cup f^{-1}(V) = X$. Finally, if $x \in f^{-1}(U) \cap f^{-1}(V)$, then $f(x) \in U \cap f(X)$ and $f(x) \in V \cap f(X)$. But $(U \cap f(X)) \cap (V \cap f(X)) = \emptyset$. Thus $f^{-1}(U) \cap f^{-1}(V) = \emptyset$ and X is disconnected. □

Corollary 5.4. *Connectedness is a topological property.*

Example. Suppose $a < b$. Then there is a homeomorphism $h \colon [0, 1] \to [a, b]$ given by $h(t) = a + (b - a)t$. Thus, every $[a, b]$ is connected.

A subspace A of a space X is disconnected when there are open sets U and V in X for which $A \cap U \neq \emptyset \neq A \cap V$, $A \subset U \cup V$, and $A \cap U \cap V = \emptyset$. Notice that $U \cap V$ can be nonempty in X, but $A \cap U \cap V = \emptyset$.

Lemma 5.5. *If $\{A_i \mid i \in J\}$ is a collection of connected subspaces of a space X with $\bigcap_{i \in J} A_i \neq \emptyset$, then $\bigcup_{i \in J} A_i$ is connected.*

Proof. Suppose U and V are open subsets of X with $\bigcup_{i \in J} A_i \subset U \cup V$ and $\bigcup_{i \in J} A_i \cap U \cap V = \emptyset$. Let $p \in \bigcap_{j \in J} A_j$; then $p \in A_j$ for all $j \in J$. Suppose that $p \in U$. Since U and V are open, $\{U \cap A_j, V \cap A_j\}$ would separate A_j if they were both nonempty. Since A_j is a connected subspace, this cannot happen, and so $A_j \subset U$. Since $j \in J$ was arbitrary, we can argue in this way to show $\bigcup A_i \subset U$, and hence $\{U, V\}$ is not a separation. $\qquad\square$

Example. Given an open interval $(a, b) \subset \mathbb{R}$, let $N > 2/(b - a)$. Then we can write $(a, b) = \bigcup_{n \geq N}[a + \frac{1}{n}, b - \frac{1}{n}]$, a union with nonempty intersection. It follows from the lemma that (a, b) is connected. Also $\mathbb{R} = \bigcup_{n > 0}[-n, n]$ and so $\mathbb{R}$ is connected.

Let us review our constructions to see how they respect connectedness. A subset A of a space X is connected if it is connected in the subspace topology. Subspaces do not generally inherit connectedness; for example, $\mathbb{R}$ is connected but $[0, 1] \cup (2, 3) \subset \mathbb{R}$ is disconnected. A quotient of a connected space, however, is connected since it is the continuous image of the connected space. How about products?

Proposition 5.6. *If X and Y are connected spaces, then $X \times Y$ is connected.*

Proof. Let x_0 and y_0 be points in X and Y, respectively. In the exercises of Chapter 4 you proved that the inclusions $j_{x_0} \colon Y \to X \times Y$, given by $j_{x_0}(y) = (x_0, y)$, and $i_{y_0} \colon X \to X \times Y$, given by $i_{y_0}(x) = (x, y_0)$, are continuous; hence $j_{x_0}(Y)$ and $i_{y_0}(X)$ are connected in $X \times Y$. Furthermore, $j_{x_0}(Y) \cap i_{y_0}(X) = (x_0, y_0)$ so $i_{y_0}(X) \cup j_{x_0}(Y)$ is connected. We express $X \times Y$ as a union of similar connected subsets:
$$X \times Y = \bigcup_{x \in X} i_{y_0}(X) \cup j_x(Y),$$
a union with intersection given by $\bigcap_{x \in X} i_{y_0}(X) \cup j_x(Y) = i_{y_0}(X)$, which is connected. By Lemma 5.5, $X \times Y$ is connected. $\qquad\square$

Example. By induction, $\mathbb{R}^n$ is connected for all n. Wrapping $\mathbb{R}$ onto S^1 by $w \colon \mathbb{R} \to S^1$, given by $w(\gamma) = (\cos(2\pi\gamma), \sin(2\pi\gamma))$, shows that S^1 is connected and so is the torus $S^1 \times S^1$. We can also prove this by arguing that $[0,1] \times [0,1]$ is connected and the torus is a quotient of $[0,1] \times [0,1]$. It also follows that S^2 is connected $S^2 \cong \Sigma S^1$, a quotient of $S^1 \times [0,1]$. By induction and Theorem 4.19, S^n is connected for all $n \geq 1$.

A characterization of the connected subspaces of $\mathbb{R}$ has some interesting corollaries.

Proposition 5.7. *If $W \subset (\mathbb{R}, usual)$ is connected, then $W = (a,b)$, $[a,b)$, $(a,b]$, or $[a,b]$ for $-\infty \leq a \leq b \leq \infty$.*

Proof. Suppose $c, d \in W$ with $c < d$. We show $[c,d] \subset W$, that is, that W is convex. (In other words, if c, d are both in W, then $(1-t)c + td \in W$ for all $0 \leq t \leq 1$.) Otherwise there exists a value r_0, with $c < r_0 < d$ and $r_0 \notin W$. Then $U = (-\infty, r_0) \cap W$, $V = W \cap (r_0, \infty)$ is a separation of W. We leave it to the reader to show that a convex subset of $\mathbb{R}$ must be an open, closed, or half-open interval.   □

Intermediate Value Theorem. *If $f: [a,b] \to \mathbb{R}$ is a continuous function and $f(a) < c < f(b)$ or $f(a) > c > f(b)$, then there is a value $x_0 \in [a,b]$ with $f(x_0) = c$.*

Proof. Since f is continuous, $f([a,b])$ is a connected subset of $\mathbb{R}$. Furthermore, this subset contains $f(a)$ and $f(b)$. By Proposition 5.7, the interval between $f(a)$ and $f(b)$, which includes c, lies in the image of $[a,b]$, and so there is a value $x_0 \in [a,b]$ with $f(x_0) = c$.   □

Corollary 5.8. *Suppose $g: S^1 \to \mathbb{R}$ is continuous. Then there is a point $x_0 \in S^1$ with $g(x_0) = g(-x_0)$.*

Proof. Define $\tilde{g} : S^1 \to \mathbb{R}$ by $\tilde{g}(x) = g(x) - g(-x)$. Wrap $[0,1]$ onto S^1 by $w(t) = (\cos(2\pi t), \sin(2\pi t))$. Then $w(0) = -w(1/2)$.

Let $F = \tilde{g} \circ w$. It follows that

$$
\begin{aligned}
F(0) &= \tilde{g}(w(0)) = g(w(0)) - g(-w(0)) \\
&= -[g(-w(0)) - g(w(0))] \\
&= -[g(w(1/2)) - g(-w(1/2))] \\
&= -F(1/2).
\end{aligned}
$$

If $F(0) > 0$, then $F(1/2) < 0$ and since F is continuous, it must take the value 0 for some t between 0 and $1/2$. Similarly for $F(0) < 0$. If $F(t) = 0$, then let $x_0 = w(t)$ and $g(x_0) = g(-x_0)$.   □

Here is a whimsical interpretation of this result: There are two antipodal points on the equator at which the temperatures are exactly the same. In later chapters we will generalize this result to continuous functions $S^n \to \mathbb{R}^n$.

It is the connectedness of the domain of a continuous real-valued function that leads to the Intermediate Value Theorem (IVT). Furthermore, the IVT can be used to prove that an odd-degree real polynomial has a real root (see the Exercises). Toward a proof of the Fundamental Theorem of Algebra, that every polynomial with complex coefficients has a complex root (see [76] and [26]), we present an argument given by Gauss, in which connectedness plays a key role. Sadly, Gauss's argument is incomplete and another deep result is needed to complete the proof (see [64]). Connectedness plays a prominent role in the argument, which illuminates the subtleness of Gauss's thinking. A complete proof of the Fundamental Theorem of Algebra, using the fundamental group, is presented in Chapter 8.

Let $p(z) = z^n + a_{n-1}z^{n-1} + \cdots + a_1 z + a_0$ be a complex monic polynomial of degree n. We begin with some estimates. We can write the complex numbers in polar form, $z = re^{i\theta}$ and $a_j = s_j e^{i\psi_j}$, and make the substitution

$$p(z) = r^n e^{ni\theta} + r^{n-1}s_{n-1}e^{(n-1)i\theta+i\psi_{n-1}} + \cdots + rs_1 e^{i\theta+i\psi_1} + s_0 e^{i\psi_0}.$$

Writing $e^{i\beta} = \cos(\beta) + i\sin(\beta)$ and $p(z) = T(z) + iU(z)$, we have

$$T(z) = r^n \cos(n\theta) + r^{n-1}s_{n-1}\cos((n-1)\theta + \psi_{n-1})$$
$$+ \cdots + rs_1 \cos(\theta + \psi_1) + s_0 \cos(\psi_0),$$
$$U(z) = r^n \sin(n\theta) + r^{n-1}s_{n-1}\sin((n-1)\theta + \psi_{n-1})$$
$$+ \cdots + rs_1 \sin(\theta + \psi_1) + s_0 \sin(\psi_0).$$

Thus a root of $p(z)$ is a complex number $z_0 = re^{i\theta_0}$ with $T(z_0) = 0 = U(z_0)$.

Suppose $S = \max\{s_{n-1}, s_{n-2}, \ldots, s_0\}$ and $R = 1 + \sqrt{2}S$. Then if $r > R$, we can write

$$0 < 1 - \frac{\sqrt{2}S}{r-1} = 1 - \sqrt{2}S\left(\frac{1}{r} + \frac{1}{r^2} + \frac{1}{r^3} + \cdots\right)$$
$$< 1 - \sqrt{2}S\left(\frac{1}{r} + \frac{1}{r^2} + \cdots + \frac{1}{r^n}\right).$$

Multiplying through by r^n, we deduce

$$0 < r^n - \sqrt{2}S(r^{n-1} + r^{n-2} + \cdots + r + 1)$$
$$\leq r^n - \sqrt{2}(s_{n-1}r^{n-1} + s_{n-2}r^{n-2} + \cdots + s_1 r + s_0).$$

The $\sqrt{2}$ factor is related to the trigonometric form of $T(z)$ and $U(z)$.

Fix a circle in the complex plane given by $z = re^{i\theta}$ for $r > R$. Denote points P_k on this circle with special values:

$$P_k = r\left(\cos\left(\frac{(2k+1)\pi}{4n}\right) + i\sin\left(\frac{(2k+1)\pi}{4n}\right)\right).$$

When we evaluate $T(P_{2k})$, the leading term is $r^n\cos(n((4k+1)\pi/4n))$ $= (-1)^k r^n(\sqrt{2}/2)$. Thus we can write $(-1)^k T(P_{2k})$ as

$$\frac{r^n}{\sqrt{2}} + (-1)^k s_{n-1}r^{n-1}\cos\left((n-1)\left(\frac{(4k+1)\pi}{4n}\right) + \psi_{n-1}\right)$$
$$+ \cdots + (-1)^k s_0\cos(\psi_0).$$

Since $(-1)^k\cos\alpha \geq -1$ for all α and $r > R$, we find that

$$(-1)^k T(P_{2k}) \geq \frac{r^n}{\sqrt{2}} - (s_{n-1}r^{n-1} + \cdots + s_1 r + s_0) > 0.$$

Similarly, in $T(P_{2k+1})$, the leading term is $(-1)^{k+1}r^n\sqrt{2}/2$ and the same estimate gives $(-1)^{k+1}T(P_{2k+1}) > 0$.

The estimates imply that the value of $T(z)$ alternates in sign at $P_0, P_1, \ldots, P_{4n-1}$. Since $T(re^{i\theta})$ varies continuously in θ, $T(z)$ has a zero between P_{2k} and P_{2k+1} for $k = 0, 1, 2, \ldots, 2n-1$. We note that these are all of the zeros of $T(z)$ on this circle. To see this, write

$$\cos\theta + i\sin\theta = \frac{1-\zeta^2}{1+\zeta^2} + i\frac{2\zeta}{1+\zeta^2}, \quad \text{where } \zeta = \tan(\theta/2).$$

Thus $T(z)$ can be written in the form

$$
r^n \left(\frac{1 - \zeta^2}{1 + \zeta^2} \right)^n + s_{n-1} \cos(\psi_{n-1}) r^{n-1} \left(\frac{1 - \zeta^2}{1 + \zeta^2} \right)^{n-1}
$$
$$
+ \cdots + s_1 \cos(\psi_1) r \left(\frac{1 - \zeta^2}{1 + \zeta^2} \right) + s_0 \cos(\psi_0),
$$

that is, $T(z) = f(\zeta)/(1 + \zeta^2)^n$, where $f(\zeta)$ is a polynomial of degree less than or equal to $2n$. Since $T(z)$ has $2n$ zeros, $f(\zeta)$ has degree $2n$ and has exactly $2n$ roots. Thus we can name the zeros of $T(z)$ on the circle of radius r by $Q_0, Q_1, \ldots, Q_{2n-1}$ with Q_k between P_{2k} and P_{2k+1}.

Let $Q_k = r e^{i\phi_k}$. Then $n\phi_k$ lies between $\frac{\pi}{4} + k\pi$ and $\frac{3\pi}{4} + k\pi$. It follows from properties of the sine function that $(-1)^k \sin(n\phi_k) \geq \sqrt{2}/2$. From this estimate we find that

$$
(-1)^k U(Q_k) \geq (-1)^k r^n \sin(n\phi_k) - s_{n-1} r^{n-1} - \cdots - s_0
$$
$$
\geq \frac{r^n}{\sqrt{2}} - s_{n-1} r^{n-1} - \cdots - s_0 > 0.
$$

Then $U(z)$ is positive at Q_{2k} and negative at Q_{2k+1} for $0 \leq k \leq n-1$, and by continuity $U(z)$ is zero between consecutive pairs of Q_j. This gives us points q_i for $i = 0, 1, \ldots, 2n - 1$ with q_i between Q_i and Q_{i+1} and $U(q_i) = 0$.

The game is clear now—a zero of $p(z)$ is a value z_0 with $T(z_0) = 0 = U(z_0)$. Gauss argued that, as the radius of the circle varied, the distinguished points Q_j and q_k would form curves. As the radius grew smaller, the points Q_k determine regions whose boundary is where $T(z) = 0$. The curve of q_j, where $U(z) = 0$, must cross some curve of Q_j's, and so give us a root of $p(z)$. The geometric properties of curves of the type given by $T(z) = 0$ and $U(z) = 0$ are needed to complete this part of the argument, and require more analysis than is appropriate here. The identification of the curves and reducing the existence of a root to the necessary intersection of curves are served up by connectedness.

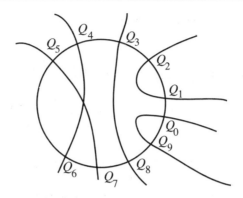

Connectedness is related to the intuitive geometric ideas of Chapter 3 by the following result.

Proposition 5.9. *If A is a connected subspace of a space X and $A \subset B \subset \operatorname{cls} A$, then B is connected.*

Proof. Suppose B has a separation $\{U \cap B, V \cap B\}$ with U, V open subsets of X. Since A is connected, either $A \subset U$ or $A \subset V$. Suppose $A \subset U$ and $x \in V$. Since V is open and $x \in V$, because $x \in B \subset \operatorname{cls} A$, we have that x is a limit point of A. Hence there is a point of A in $U \cap V$ and so $x \in B \cap U \cap V$. This contradicts the assumption that $\{U \cap B, V \cap B\}$ is a separation. Thus B is connected.   □

Some wild connected spaces can be constructed from this proposition.

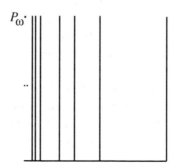

Let $P_\omega = (0,1) \in \mathbb{R}^2$ and let X be the subspace of $\mathbb{R}^2$ given by

$$X = \{P_\omega\} \cup \left((0,1] \times \{0\}\right) \cup \left(\bigcup_{n=1}^\infty \left\{\frac{1}{n}\right\} \times [0,1]\right).$$

We call X the *deleted comb space* [61]. The spokes together with the base form a connected subspace of X. The stray point P_ω is the limit point of the sequence given by the tops of the spokes, $\{(1/n, 1)\}$. So X lies between the connected space of the spokes and base and its closure. Hence X is connected.

Connectedness determines an equivalence relation on a space X: $x \sim y$ if there is a connected subset A of X with $x, y \in A$. (Can you prove that this is an equivalence relation?) An equivalence class $[x]$ under this relation is called a **connected component** of X. The equivalence classes satisfy the property that if $x \in [x]$, then $[x]$ is the union of all connected subsets of X containing x and so it follows from Lemma 5.5 that $[x]$ is the largest connected subset containing x. Since $[x] \subseteq \mathrm{cls}[x]$, it follows from Proposition 5.9 that $\mathrm{cls}[x]$ is also connected and hence $[x] = \mathrm{cls}[x]$ and connected components are closed.

Because the connected components partition a space, and each is closed, then each is also open if there are only finitely many connected components. By way of contrast with the case of finitely many components, the connected components of $\mathbb{Q} \subset \mathbb{R}$ are the points themselves—closed but not open.

Proposition 5.10. *The cardinality of the set of connected components of a space X is a topological invariant.*

Proof. We show that if $[x]$ is a component of X and $h\colon X \to Y$ is a homeomorphism, then $h([x])$ is a component of Y. By Theorem 5.3, $h([x])$ is connected and $h([x]) \subset [h(x)]$. By a symmetric argument, $h^{-1}([h(x)]) \subset [x]$. Thus $[h(x)] \subset h([x])$ and so $h([x]) = [h(x)]$. Since h maps components to components, h induces a one-one correspondence between connected components. $\square$

We have developed enough topology to handle a case of our main goal. Connectedness allows us to distinguish between $\mathbb{R}$ and $\mathbb{R}^n$ for $n \geq 2$.

Invariance of Dimension for $(1, n)$. $\mathbb{R}$ *is not homeomorphic to* $\mathbb{R}^n$, *for* $n > 1$.

We first make a useful observation.

Lemma 5.11. *If* $f : X \to Y$ *is a homeomorphism and* $x \in X$, *then* f *induces a homeomorphism between* $X - \{x\}$ *and* $Y - \{f(x)\}$.

Proof. The restriction $f| : X - \{x\} \to Y - \{f(x)\}$ of f to $X - \{x\}$ is a one-one correspondence between $X - \{x\}$ and $Y - \{f(x)\}$. Each subset is endowed with the subspace topology and $f|$ is continuous because an open set in $Y - \{f(x)\}$ is the intersection of an open set V in Y with the complement of $\{f(x)\}$. The inverse image is the intersection of $f^{-1}(V)$ and the complement of $\{x\}$, an open set in $X - \{x\}$. The inverse of $f|$ is similarly seen to be continuous. $\square$

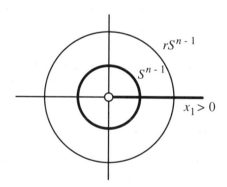

Proof of Invariance of Dimension for $(1, n)$. Suppose we had a homeomorphism $h : \mathbb{R} \to \mathbb{R}^n$. By composing with a translation we arrange that $h(0) = \mathbf{0} = (0, 0, \ldots, 0) \in \mathbb{R}^n$. By Lemma 5.11, we consider the homeomorphism $h| : \mathbb{R} - \{0\} \to \mathbb{R}^n - \{\mathbf{0}\}$. But $\mathbb{R} - \{0\}$ has two connected components. To demonstrate invariance of dimension in this case we show for $n > 1$ that $\mathbb{R}^n - \{\mathbf{0}\}$ has only one component. Fix the connected subset of $\mathbb{R}^n - \{\mathbf{0}\}$ given by

$$Y = \{(x_1, 0, \ldots, 0) \mid x_1 > 0\}.$$

This is an open ray, which we know to be connected. We can express $\mathbb{R}^n - \{0\}$ as a union:

$$\mathbb{R}^n - \{0\} = \bigcup_{r>0} rS^{n-1} \cup Y,$$

where $rS^{n-1} = \{(a_1, \ldots, a_n) \in \mathbb{R}^n \mid a_1^2 + \cdots + a_n^2 = r^2\}$. Each subset in the union is connected since it is the union of a homeomorphic copy of S^{n-1} and Y with nonempty intersection. The intersection of all of the sets in the union is Y and so, by Lemma 5.5, $\mathbb{R}^n - \{0\}$ is connected and thus has only one component. $\square$

Path-connectedness

A more natural formulation of connection is given by the following notion.

Definition 5.12. A space X is **path-connected** if, for any $x, y \in X$, there is a continuous function $\lambda \colon [0,1] \to X$ with $\lambda(0) = x$, $\lambda(1) = y$. Such a function λ is called a **path** joining x to y in X.

The connectedness of $[0,1]$ plays a role in relating connectedness with path-connectedness.

Proposition 5.13. *If X is path-connected, then it is connected.*

Proof. Suppose X is disconnected and $\{U, V\}$ is a separation. Since $U \neq \emptyset \neq V$, there are points $x \in U$ and $y \in V$. If X is path-connected, there is a path $\lambda \colon [0,1] \to X$ with $\lambda(0) = x$, $\lambda(1) = y$, and λ continuous. But then $\{\lambda^{-1}(U), \lambda^{-1}(V)\}$ would separate $[0,1]$, a connected space. This contradiction implies that X is connected. $\square$

Connectedness and path-connectedness are not equivalent. We saw that the *deleted comb space* is connected but it is not path-connected. Suppose there is a path $\lambda \colon [0,1] \to X$ with $\lambda(0) = (1,0)$ and $\lambda(1) = (0,1) = P_\omega$. The subset $\lambda^{-1}(\{P_\omega\})$ is closed in $[0,1]$ because X is Hausdorff and λ is continuous. We will show that it is also open. Consider $V = B(P_\omega, \epsilon) \cap X$ for $\epsilon = 1/k > 0$ and $k > 1$. Then $\lambda^{-1}(V)$ is nonempty and open in $[0,1]$, so for $x_0 \in \lambda^{-1}(V)$, there exists $\delta > 0$ with $(x_0 - \delta, x_0 + \delta) \cap [0,1] \subset \lambda^{-1}(V)$. We claim that $(x_0 - \delta, x_0 + \delta) \subset \lambda^{-1}(\{P_\omega\})$. Suppose not and T is such that $\lambda(T) = (\frac{1}{n}, s)$ for some $n > k$. Let $W_1 = (-\infty, r) \times \mathbb{R}$ and $W_2 = (r, \infty) \times \mathbb{R}$ for $1/(n+1) < r < 1/n$. Then $\{W_1 \cap \lambda((x_0 - \delta, x_0 + \delta)), W_2 \cap \lambda((x_0 - \delta, x, +\delta))\}$ separates the image $\lambda((x_0 - \delta, x_0 + \delta))$ of a connected space under a continuous mapping, and this is a contradiction. It follows that no such value of T exists. Since $\lambda^{-1}(\{P_\omega\}) \subset [0,1]$ is both open and closed, λ is a constant path with image P_ω.

By analogy with the property of connectedness, we have the following results.

Theorem 5.14. *If X is path-connected and $f \colon X \to Y$ is continuous, then $f(X) \subset Y$ is path-connected.*

Proof. Let $f(x), f(y) \in f(X)$. There is a path $\lambda \colon [0,1] \to X$ joining x and y. Then $f \circ \lambda$ is a path joining $f(x)$ and $f(y)$. $\qquad \square$

Corollary 5.15. *Path-connectedness is a topological property.*

Lemma 5.16. *If $\{A_i \mid i \in J\}$ is a collection of path-connected subsets of a space X and $\bigcap_{i \in J} A_i \neq \emptyset$, then $\bigcup_{i \in J} A_i$ is path-connected.*

Proof. Suppose $x, y \in \bigcup_{i \in J} A_i$ and $z \in \bigcap_{i \in J} A_i$. Then, for some i_1 and $i_2 \in J$, we have $x \in A_{i_1}$, $y \in A_{i_2}$, both subsets path-connected. Then there are paths $\lambda_1 \colon [0,1] \to A_{i_1}$, with $\lambda_1(0) = x$, $\lambda_1(1) = z$, and $\lambda_2 \colon [0,1] \to A_{i_2}$, with $\lambda_2(0) = z$, $\lambda_2(1) = y$. Define the path $\lambda_1 * \lambda_2$ by

$$\lambda_1 * \lambda_2(t) = \begin{cases} \lambda_1(2t), & 0 \leq t \leq \frac{1}{2}, \\ \lambda_2(2t - 1), & \frac{1}{2} \leq t \leq 1. \end{cases}$$

By Theorem 4.4, the path $\lambda_1 * \lambda_2$ is continuous. Furthermore, $\lambda_1 * \lambda_2$ joins x to y and so $\bigcup_{i \in J} A_i$ is path-connected. $\qquad \square$

By Proposition 5.7, the connected subsets of $\mathbb{R}$ are intervals. If $r, s \in (a, b)$, then the path $t \mapsto (1-t)r + ts$ joins r to s in (a, b). Thus, the connected subspaces of $\mathbb{R}$ are path-connected.

As is the case for connectedness, path-connectedness of subspaces of a path-connected space is unpredictable. However, by Theorem 5.14 quotients of path-connected spaces are connected. We consider products.

Proposition 5.17. *If X and Y are path-connected, then so is $X \times Y$.*

Proof. Let (x, y) and (x', y') be points in $X \times Y$. Since X and Y are path-connected, there are paths $\lambda \colon [0, 1] \to X$ and $\lambda' \colon [0, 1] \to Y$ with $\lambda(0) = x$, $\lambda(1) = x'$, $\lambda'(0) = y$, and $\lambda'(1) = y'$. Consider $\lambda \times \lambda' \colon [0, 1] \to X \times Y$ given by

$$(\lambda \times \lambda')(t) = (\lambda(t), \lambda'(t)).$$

By Proposition 4.10, $\lambda \times \lambda'$ is continuous with $\lambda \times \lambda'(0) = (x, y)$ and $\lambda \times \lambda'(1) = (x', y')$ as required. So $X \times Y$ is path-connected. $\qquad\Box$

This shows, by induction, that $\mathbb{R}^n$ is path-connected for all n. Together with the remark about quotients, spaces such as S^{n-1}, $S^1 \times S^1$, and $\mathbb{R}P^2$ are all path-connected.

Paths lead to another relation on a space X: we write $x \approx y$ if there is a path $\lambda \colon [0, 1] \to X$ with $\lambda(0) = x$ and $\lambda(1) = y$. The constant path $c_{x_0} \colon [0, 1] \to X$, given by $c_{x_0}(t) = x_0$, is continuous and so, for all $x_0 \in X$, $x_0 \approx x_0$. If $x \approx y$, then there is a path λ joining x to y. Consider the mapping $\lambda^{-1}(t) = \lambda(1 - t)$. Then λ^{-1} is continuous and determines a path joining y to x. Thus $y \approx x$. Finally, if $x \approx y$ and $y \approx z$, then if λ_1 joins x to y and λ_2 joins y to z, then $\lambda_1 * \lambda_2$ joins x to z, and so the relation $\approx$ is an equivalence relation.

We define a **path component** to be an equivalence class under the relation $\approx$. A space is path-connected if and only if it has only one path component. Since each path component $[x]$ is path-connected we know that for $f \colon X \to Y$ a continuous function, $f([x]) \subset [f(x)]$, since the image of a path-connected subspace is path-connected. We extend this fact a little further as follows.

Definition 5.18. The **set of path components** $\pi_0(X)$ is the set of equivalence classes under the relation $\approx$. If $f \colon X \to Y$ is a continuous function, then f induces a well-defined mapping $\pi_0(f) \colon \pi_0(X) \to \pi_0(Y)$, given by $\pi_0(f)([x]) = [f(x)]$.

We note that the association $X \mapsto \pi_0(X)$ and $f \mapsto \pi_0(f)$ satisfies the following basic properties: (1) If id $\colon X \to X$ is the identity mapping, then $\pi_0(\mathrm{id}) \colon \pi_0(X) \to \pi_0(X)$ is the identity mapping;

(2) If $f\colon X \to Y$ and $g\colon Y \to Z$ are continuous mappings, then $\pi_0(g \circ f) = \pi_0(g) \circ \pi_0(f)\colon \pi_0(X) \to \pi_0(Z)$. These properties are shared with several constructions to come and they came to be identified as the *functoriality* of π_0 [**22**]. The alert reader will recognize functoriality at work in later chapters.

As with connected components, we ask when path components are open or closed. The deleted comb space, however, indicates that we cannot expect much of closure.

Definition 5.19. A space X is **locally path-connected** if, for every $x \in X$ and $x \in U$ an open set in X, there is an open set $V \subset X$ with $x \in V \subset U$ and V path-connected.

Proposition 5.20. *If X is locally path-connected, then path components of X are open.*

Proof. Let $y \in [x]$, a path component of X. Take any open set containing y and there is a path-connected open set V_y with $y \in V_y$. Since every point in V_y is related to y and y is related to x, we get that $V_y \subset [x]$. Thus $[x] = \bigcup_{y \in [x]} V_y$ and $[x]$ is open. □

We see how this can work together with connectedness to obtain path-connectedness.

Corollary 5.21. *If X is connected and locally path-connected, then it is path-connected.*

Proof. Suppose X has more than one path component. Choose one component $[x] = U$, which is open in X. The union of the rest of the components we denote by V, which is also open in X. Then $U \cup V = X$ and $U \cap V = \emptyset$, and so X is disconnected, a contradiction. Hence X has only one path component. □

It follows that deleted comb space is not even locally path-connected. (This can also be proved directly.)

Exercises

    1. Prove that any infinite set X with the finite-complement topology is connected. Is the space $(\mathbb{R}, \text{half-open})$ connected?

2. A subset $K \subset \mathbb{R}$ is convex if for any $c, d \in K$, the set $[c, d] = \{c(1 - t) + dt \mid 0 \leq t \leq 1\}$ is contained in K. Show that a convex subset of $\mathbb{R}$ is an open, closed, or half-open interval.

3. $\ldots$, *the hip bone's connected to the thigh bone, and the thigh bone's connected to the knee bone, and the* $\ldots$. Let's prove a proposition that shows that the skeleton should be connected as in the song. Suppose we have a sequence of connected subspaces $\{X_i \mid i = 1, 2, 3, \ldots\}$ of a given space X. Suppose further that $X_i \cap X_{i+1} \neq \emptyset$ for all i. Show that the union $\bigcup_{i=1}^{\infty} X_i$ is connected. (Hint: Consider the sequence of subspaces $Y_j = X_1 \cup X_2 \cup \cdots \cup X_j$ for $j \geq 1$. Are these connected? What is their intersection? What is their union?)

4. Suppose we have a collection of nonempty connected spaces, $\{X_j \mid j \in J\}$. Does it follow that the product $\prod_{j \in J} X_j$ is connected?

5. One of the easier parts of the Fundamental Theorem of Algebra is the fact that an odd degree polynomial $p(x)$ has at least one real root. Notice that such a polynomial is a continuous function $p \colon \mathbb{R} \to \mathbb{R}$. The theorem follows by showing that there is a real number b with $p(b) > 0$ and $p(-b) < 0$, and using the Intermediate Value Theorem.

   Let $p(x) = x^n + a_{n-1}x^{n-1} + \cdots + a_1 x + a_0$ with n odd. Write $p(x) = x^n q(x)$ for the function $q(x)$ that will be the sum of the coefficients of $p(x)$ over powers of x. Estimate $|q(x) - 1|$ and show that it is less than or equal to $A/|x|$, where $A = |a_{n-1}| + \cdots + |a_1| + |a_0|$ for $|x| \geq 1$. Letting $|b| > \max\{1, 2A\}$ we get $|q(b) - 1| < \frac{1}{2}$ or $q(b) > 0$ and $q(-b) > 0$. Show that this implies that there is a zero of $p(x)$ between $-|b|$ and $|b|$.

6. Suppose that the space X can be written as a product $X = Y_1 \times Y_2$. Determine the relationship between $\pi_0(X)$ and $\pi_0(Y_1)$ and $\pi_0(Y_2)$. Suppose that G is a topological group. Show that $\pi_0(G)$ is also a group.

Chapter 6

Compactness

... compact sets play the same role in the theory of abstract sets as the notion of limit sets do in the theory of point sets.

MAURICE FRÉCHET, 1906

Compactness is one of the most useful topological properties in analysis, although, at first meeting its definition seems somewhat strange. To motivate the notion of a compact space, consider the properties of a *finite* subset $S \subset X$ of a topological space X. Among the consequences of finiteness are the following:

i) Any continuous function $f \colon X \to \mathbb{R}$, when restricted to S, has a maximum and a minimum.

ii) Any collection of open subsets of X whose union contains S has a finite subcollection whose union contains S.

iii) Any sequence of points $\{x_i\}$ satisfying $x_i \in S$ for all i has a convergent subsequence.

Compactness extends these properties to other subsets of a space X, using the topology to achieve what finiteness guarantees. The development in this chapter runs parallel to that of Chapter 5 on connectedness.

Definition 6.1. Given a topological space X and a subset $K \subset X$, a collection of subsets $\{C_i \subset X \mid i \in J\}$ is a **cover** of K if $K \subset \bigcup_{i \in J} C_i$. A cover is an **open cover** if every C_i is open in X. The cover $\{C_i \mid i \in J\}$ of K has a **finite subcover** if there are members of the collection $C_{i_1}, \ldots, C_{i_n}$ with $K \subset C_{i_1} \cup \cdots \cup C_{i_n}$. A subset $K \subset X$ is **compact** if any open cover of K has a finite subcover.

Examples. Any finite subset of a topological space is compact. The space $(\mathbb{R}, \text{usual})$ is not compact since the open cover $\{(-n, n) \mid n = 1, 2, \ldots\}$ has no finite subcover. Notice that if K is a subset of $\mathbb{R}^n$ and K is compact, it is *bounded*, that is, $K \subset B(\vec{0}, M)$ for some $M > 0$. This follows since $\{B(\vec{0}, N) \mid N = 1, 2, \ldots\}$ is an open cover of $\mathbb{R}^n$ and hence of K. Since $B(\vec{0}, N_1) \subset B(\vec{0}, N_2)$ for $N_1 \leq N_2$, a finite subcover is contained in a single open ball and so K is bounded. The canonical example of a compact space is the unit interval $[0, 1] \subset \mathbb{R}$.

The Heine-Borel Theorem. *The closed interval $[0, 1]$ is a compact subspace of $(\mathbb{R}, usual)$.*

Proof. Suppose $\{U_i \mid i \in J\}$ is an open cover of $[0, 1]$. Define $T = \{x \in [0, 1] \mid [0, x]$ has a finite subcover from $\{U_i\}\}$. Certainly $0 \in T$ since $0 \in \bigcup U_i$ and so in some U_j. We show $1 \in T$. Since every element of T is less than or equal to 1, T has a least upper bound s. Since $\{U_i\}$ is a cover of $[0, 1]$, $s \in U_j$ for some $j \in J$. Since U_j is open, $(s - \epsilon, s + \epsilon) \subset U_j$ for some $\epsilon > 0$. Because s is a least upper bound, $s - \delta \in T$ for some $0 < \delta < \epsilon$ and so $[0, s - \delta]$ has a finite subcover. It follows that $[0, s]$ has a finite subcover by simply adding U_j to the finite subcover of $[0, s - \delta]$. If $s < 1$, then there is an $\eta > 0$ with $s + \eta \in (s - \epsilon, s + \epsilon) \cap [0, 1]$, and so $s + \eta \in T$, which contradicts the fact that s is a least upper bound. Hence $s = 1$. $\qquad\square$

Is compactness a topological property? We prove a result analogous to Theorem 5.2 for connectedness.

Theorem 6.2. *If $f \colon X \to Y$ is a continuous function and X is compact, then $f(X) \subset Y$ is compact.*

Proof. Suppose $\{U_i \mid i \in J\}$ is an open cover of $f(X)$ in Y. Then $\{f^{-1}(U_i) \mid i \in J\}$ is an open cover of X. Since X is compact, there is a finite subcover $\{f^{-1}(U_{i_1}), \ldots, f^{-1}(U_{i_n})\}$. Then $X = f^{-1}(U_{i_1}) \cup \cdots \cup f^{-1}(U_{i_n})$ and so $f(X) \subset U_{i_1} \cup \cdots \cup U_{i_n}$. □

It follows immediately that compactness is a topological property. The closed interval $[a, b] \subset (\mathbb{R}, \text{usual})$ is compact for $a < b$. Since S^1 is the continuous image of $[0, 1]$, S^1 is compact. Notice that compactness distinguishes the open and closed intervals in $\mathbb{R}$. Since $(0, 1)$ is homeomorphic to $\mathbb{R}$ and $\mathbb{R}$ is not compact, $(0, 1) \ncong [0, 1]$. Since $(0, 1) \subset [0, 1]$, arbitrary subspaces of compact spaces need not be compact. However, compactness is inherited by closed subsets.

Proposition 6.3. *If X is a compact space and $K \subset X$ is a closed subset, then K is compact.*

Proof. If $\{U_i \mid i \in J\}$ is an open cover of K, we can take the collection $\{X - K\} \cup \{U_i \mid i \in J\}$ as an open cover of X. Since X is compact, the collection has a finite subcover, namely $\{X - K, U_{i_1}, \ldots, U_{i_n}\}$. Leaving out $X - K$, we get $\{U_{i_1}, \ldots, U_{i_n}\}$, a finite subcover of K. □

A partial converse holds for Hausdorff spaces.

Proposition 6.4. *If X is Hausdorff and $K \subset X$ is compact, then K is closed in X.*

Proof. We show $X - K$ is open. Take $x \in X - K$. By the Hausdorff condition, for each $y \in K$ there are open sets U_y, V_y with $x \in U_y$, $y \in V_y$, and $U_y \cap V_y = \emptyset$. Then $\{V_y \mid y \in K\}$ is an open cover of K. Since K is compact, there is a finite subcover $\{V_{y_1}, V_{y_2}, \ldots, V_{y_n}\}$. Take the associated open sets $U_{y_1}, \ldots, U_{y_n}$ and define $U_x = U_{y_1} \cap \cdots \cap U_{y_n}$. Since $U_{y_i} \cap V_{y_i} = \emptyset$, U_x doesn't meet $V_{y_1} \cup \cdots \cup V_{y_n} \supset K$. So $U_x \subset X - K$. Furthermore, $x \in U_x$ and U_x is open. Construct U_x for every point x in $X - K$, and the union of these open sets U_x is $X - K$ and K is closed. □

Corollary 6.5. *If $K \subset \mathbb{R}^n$ is compact, K is closed and bounded.*

Quotient spaces of compact spaces are seen to be compact by Theorem 6.2. The converse of Corollary 6.5 will follow from a consideration of finite products.

Proposition 6.6. *If X and Y are compact spaces, then $X \times Y$ is compact.*

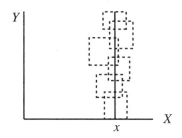

Proof. Suppose $\{U_i \mid i \in J\}$ is an open cover of $X \times Y$. From the definition of the product topology, each $U_i = \bigcup_{j \in A_i} V_{ij} \times W_{ij}$, where V_{ij} is open in X, W_{ij} is open in Y, and A_i is an indexing set. Consider the associated open cover $\{V_{ij} \times W_{ij} \mid i \in J, j \in A_i\}$ by basic open sets. If we can manufacture a finite subcover from this collection, we can just take the U_i in which each basic open set sits to get a finite subcover of $X \times Y$.

To each $x \in X$ consider the subspace $\{x\} \times Y \subset X \times Y$. This subspace is homeomorphic to Y and hence is compact. Furthermore $\{V_{ij} \times W_{ij} \mid i \in J, j \in A_i\}$ covers $\{x\} \times Y$ and so there is a finite subcover $V_1^x \times W_1^x, \ldots, V_e^x \times W_e^x$ of $\{x\} \times Y$. Let $V^x = V_1^x \cap \cdots \cap V_e^x$. Since $x \in V^x$, it is a nonempty open set. Construct V^x for each $x \in X$ and the collection $\{V^x \mid x \in X\}$ is an open cover of X. Since X is compact, there is a finite subcover $V^{x_1}, \ldots, V^{x_n}$. Hence each $x \in X$ appears in some V^{x_i}. If $y \in Y$, then $(x, y) \in V_j^{x_i} \times W_j^{x_i}$ for some $W_j^{x_i}$ since $x \in V_1^{x_i} \cap \cdots \cap V_{e_i}^{x_i}$ and $V_1^{x_i} \times W_1^{x_i}, \ldots, V_{e_i}^{x_i} \times W_{e_i}^{x_i}$ covers $\{x\} \times Y$. Hence $\{V_j^{x_i} \times W_j^{x_i} \mid i = 1, \ldots, n, j = 1, \ldots, e_i\}$ is a finite subcover of $X \times Y$. The associated choices of U_i's give the finite subcover we seek. $\square$

By induction, any finite product of compact spaces is compact. Since $[0, 1] \times [0, 1]$ is compact, so are the quotients given by the torus, Möbius band, and projective plane. We can now prove the converse of Corollary 6.5.

Corollary 6.7. *If $K \subset \mathbb{R}^n$, then K is compact if and only if K is closed and bounded.*

Proof. A bounded subset of $\mathbb{R}^n$ is contained in some product of closed intervals $[a_1, b_1] \times \cdots \times [a_n, b_n]$. The product is compact, and K is a closed subset of $[a_1, b_1] \times \cdots \times [a_n, b_n]$. By Proposition 6.3, K is compact. $\square$

We can add the spheres $S^{n-1} \subset \mathbb{R}^n$ to the list of compact spaces—each is bounded by definition and closed because $S^{n-1} = f^{-1}(\{1\})$, where $f \colon \mathbb{R}^n \to \mathbb{R}$ is the continuous function $f(x_1, \ldots, x_n) = x_1^2 + \cdots + x_n^2$. The characterization of compact subsets of $\mathbb{R}$ leads to the following familiar result.

The Extreme Value Theorem. *If $f \colon X \to \mathbb{R}$ is a continuous function and X is compact, then there are points x_m, $x_M \in X$ with $f(x_m) \leq f(x) \leq f(x_M)$ for all $x \in X$.*

Proof. By Proposition 6.2, $f(X)$ is a compact subset of $\mathbb{R}$ and so $f(X)$ is closed and bounded. The boundedness implies that the greatest lower bound of $f(X)$ and the least upper bound of $f(X)$ exist. Since $f(X)$ is closed, the values $\mathrm{glb}\, f(X)$ and $\mathrm{lub}\, f(X)$ are in $f(X)$ (Can you prove this?) and so $\mathrm{glb}\, f(X) = f(x_m)$ for some $x_m \in X$; also $\mathrm{lub}\, f(X) = f(x_M)$ for some $x_M \in X$. It follows that $f(x_m) \leq f(x) \leq f(x_M)$ for all $x \in X$. $\square$

The reader might enjoy deriving the whole of the single variable calculus armed with the Intermediate Value Theorem and the Extreme Value Theorem.

Infinite products of compact spaces are covered by the following powerful theorem which turns out to be equivalent to the Axiom of Choice in set theory [**44**]. We refer the reader to [**61**] for a proof.

Tychonoff's Theorem. *If $\{X_i \mid i \in J\}$ is a collection of nonempty compact spaces, then, with the product topology, $\prod_{i \in J} X_i$ is compact.*

Infinite products give a different structure in which to consider families of functions with certain properties as subspaces of a product. General products also provide spaces in which there is a lot of room for embedding classes of spaces as subspaces of a product.

Compact spaces enjoy some other interesting properties:

Proposition 6.8. *If $R = \{x_\alpha \mid \alpha \in J\}$ is an infinite subset of a compact space X, then R has a limit point.*

Proof. Suppose R has no limit points. The absence of limit points implies that, for every $x \in X$, there is an open set U^x with $x \in U^x$ for which, if $x \in R$, then $U^x \cap R = \{x\}$ and if $x \notin R$, then $U^x \cap R = \emptyset$. The collection $\{U^x \mid x \in X\}$ is an open cover of X, which is compact, and so it has a finite subcover $U^{x_1}, \ldots, U^{x_n}$. Since each U^{x_i} contains at most one element in $\{x_\alpha \mid \alpha \in J\}$, the set R is finite. $\qquad\square$

The property that an infinite subset must have a limit point is sometimes called the *Bolzano-Weierstrass property* [**61**]. The proposition gives a sufficient test for noncompactness: Find a sequence without a limit point. For example, if we give $\prod_{i=1}^\infty [0,1]$, the countable product of $[0,1]$ with itself, the box topology, then the set $\{(x_{n,i}) \in \prod_{i=1}^\infty \mid n = 1, 2, \ldots\}$, given by $x_{n,i} = 1$ when $n \neq i$ and $x_{n,i} = 1/n$ if $n = i$, has no limit point. (Can you prove it?)

Compactness provides a simple condition for a mapping to be a homeomorphism.

Proposition 6.9. *If $f \colon X \to Y$ is continuous, one-one, and onto, X is compact, and Y is Hausdorff, then f is a homeomorphism.*

Proof. We show that f^{-1} is continuous by showing that f is closed (that is, $f(K)$ is closed whenever K is closed). If $K \subset X$ is closed, then it is compact. It follows that $f(K)$ is compact in Y and so $f(K)$ is closed because Y is Hausdorff. $\qquad\square$

Proposition 6.9 can make the comparison of quotient spaces and other spaces easier. For example, suppose X is a compact space with an equivalence relation $\sim$ on it and $\pi \colon X \to [X]$ is a quotient mapping. Given a mapping $f \colon X \to Y$ for which $x \sim x'$ implies $f(x) = f(x')$, we get an induced mapping $\hat{f} \colon [X] \to Y$ that may be one-one, onto, and continuous. If Y is Hausdorff, we obtain that $[X]$ is homeomorphic to Y.

What about compact metric spaces? The **diameter** of a subset A of a metric space X is defined by $\operatorname{diam} A = \sup\{d(x,y) \mid x, y \in A\}$.

Lebesgue's Lemma. *Let X be a compact metric space and $\{U_i \mid i \in J\}$ an open cover. Then there is a real number $\delta > 0$ (**the Lebesgue number**) such that any subset of X of diameter less than δ is contained in some U_i.*

Proof. In the exercises to Chapter 3 we defined the continuous function $d(-, A) \colon X \to \mathbb{R}$ by $d(x, A) = \inf\{d(x, a) \mid x \in A\}$. In addition, if A is closed, then $d(x, A) > 0$ for $x \notin A$. Given an open cover $\{U_i \mid i \in J\}$ of the compact space X, there is a finite subcover $\{U_{i_1}, \ldots, U_{i_n}\}$. Define $\varphi_j(x) = d(x, X - U_{i_j})$ for $j = 1, 2, \ldots, n$ and let $\varphi(x) = \max\{\varphi_1(x), \ldots, \varphi_n(x)\}$. Since each $x \in X$ lies in some $U_{i_j}, \varphi(x) \geq \varphi_j(x) > 0$. Furthermore, φ is continuous so $\varphi(X) \subset \mathbb{R}$ is compact, and $0 \notin \varphi(X)$. Let $\delta = \min\{\varphi(x) \mid x \in X\} > 0$. For any $x \in X$, consider $B(x, \delta) \subset X$. We know $\varphi(x) = \varphi_j(x)$ for some j. For that j, $d(x, X - U_{i_j}) \geq \delta$, which implies $B(x, \delta) \subset U_{i_j}$. $\square$

The Lebesgue Lemma is also known as the *Pflastersatz* [5] (imagine plasters covering a space) and it will play a key role in later chapters.

By analogy with connectedness and path-connectedness we introduce the local version of compactness.

Definition 6.10. A space X is **locally compact** if for any $x \in U \subset X$, where U is an open set, there is an open set V satisfying $x \in V \subset \operatorname{cls} V \subset U$ with $\operatorname{cls} V$ compact.

Examples. For all n, the space $\mathbb{R}^n$ is locally compact since each $\operatorname{cls} B(\vec{x}, \epsilon)$ is compact (being closed and bounded). The countable product of copies of $\mathbb{R}$, $\prod_{i=1}^{\infty} \mathbb{R}$, in the product topology, however, is not locally compact. To see this consider any open set of the form

$$U = (a_1, b_1) \times (a_2, b_2) \times \cdots \times (a_n, b_n) \times \mathbb{R} \times \mathbb{R} \times \cdots$$

whose closure is $[a_1, b_1] \times [a_2, b_2] \times \cdots \times [a_n, b_n] \times \mathbb{R} \times \mathbb{R} \times \cdots$. This set is not compact because there is plenty of room for infinite sets to float off without limit points. Thus local compactness distinguishes finite and infinite products of $\mathbb{R}$, a partial result toward the topological invariance of dimension.

In the presence of local compactness and a little more, we can make a noncompact space into a compact one.

Definition 6.11. Let X be a locally compact, Hausdorff space. Adjoin a point not in X, denoted by ∞, to form $Y = X \cup \{\infty\}$. Topologize Y by two kinds of open sets:

(1) $U \subset X \subset Y$ and U is open in X.

(2) $Y - K$, where K is compact in X.

The space Y with this topology is called the **one-point compactification** of X.

The one-point compactification was introduced by Alexandroff [4] and is also called the *Alexandroff compactification*. We verify that we have a topology on Y as follows: For finite intersections there are three cases: We only need to consider the case of two open sets. (1) If V_1 and V_2 are both open subsets of X, then $V_1 \cap V_2$ is also an open subset of X and hence of Y. (2) If both V_1 and V_2 have the form $Y - K_1$ and $Y - K_2$, where K_1 and K_2 are compact in X, then $(Y - K_1) \cap (Y - K_2) = Y - (K_1 \cup K_2)$ and $K_1 \cup K_2$ is compact in X, so $V_1 \cap V_2$ is open in Y. (3) If V_1 is an open subset of X and $V_2 = Y - K_2$ for K_2 compact in X, then $V_1 \cap V_2 = V_1 \cap (Y - K_2) = V_1 \cap (X - K_2)$ since $V_1 \subset X$. Since $X - K_2$ is open in X, the intersection $V_1 \cap V_2$ is open in Y.

For arbitrary unions there are three similar cases. If $\{V_\beta \mid \beta \in I\}$ is a collection of open sets, then $\bigcup V_\beta$ is certainly open when $V_\beta \subset X$ for all β. If $V_\beta = Y - K_\beta$ for all β, then DeMorgan's law gives

$$\bigcup_\beta (Y - K_\beta) = Y - \bigcap_\beta K_\beta$$

and $\bigcap_\beta K_\beta$ is compact. Finally, if the V_β are of different types, the set-theoretic fact $U \cup (Y - K) = Y - (K - U)$ together with the fact that if K is compact, then, since $K - U$ is a closed subset of K, so is $K - U$ compact. Thus the union of the V_β is open in Y.

Theorem 6.12. *If X is locally compact and Hausdorff, X is not compact, and $Y = X \cup \{\infty\}$ is the one-point compactification, then Y is a compact Hausdorff space, X is a subspace of Y, and* cls $X = Y$.

Proof. We first show Y is compact. If $\{V_i \mid i \in J\}$ is an open cover of Y, then $\infty \in V_{j_0}$ for some $j_0 \in J$ and $V_{j_0} = Y - K_{j_0}$ for K_{j_0} compact in X. Since any open set in Y satisfies the property that

$V_i \cap X$ is open in X, the collection $\{V_i \cap X \mid i \in J, i \neq j_0\}$ is an open cover of K_{j_0} and so there is a finite subcover $V_1 \cap X, \ldots, V_n \cap X$ of X. Then $\{V_{j_0}, V_1, \ldots, V_n\}$ is a finite subcover of Y.

Next we show Y is Hausdorff. The important case to check is a separation of $x \in X$ and ∞. Since X is locally compact and X is open in X, there is an open set $V \subset X$ with $x \in V$ and $\operatorname{cls} V$ compact. Take V to contain x and $Y - \operatorname{cls} V$ to contain ∞. Since X is not compact, $\operatorname{cls} V \neq X$.

Notice that the inclusion $i \colon X \to Y$ is continuous since $i^{-1}(Y - K) = X - K$ and K is closed in the Hausdorff space X. Furthermore, i is an open map so X is homeomorphic to $Y - \{\infty\}$. To prove that $\operatorname{cls} X = Y$, check that ∞ is a limit point of X: if $\infty \in Y - K$, since X is not compact, $K \neq X$ so there is a point of X in $Y - K$ not equal to ∞. $\qquad\square$

Example. Stereographic projection of the sphere S^2 minus the North Pole onto $\mathbb{R}^2$ shows that the one-point compactification of $\mathbb{R}^2$ is homeomorphic to S^2. Recall that stereographic projection is defined as the mapping from S^2 minus the North Pole to the plane tangent to the South Pole by joining a point on the sphere to the North Pole and then extending this line to meet the tangent plane. This mapping has wonderful properties ([**59**]) and gives the homeomorphism between $\mathbb{R}^2 \cup \{\infty\}$ and S^2. More generally, $\mathbb{R}^n \cup \{\infty\} \cong S^n$.

Compactness may be used to define a topology on $\operatorname{Hom}(X, Y) = \{f \colon X \to Y$ such that f is continuous$\}$. There are many possible choices, some dependent on the topologies of X and Y (for example, for metric spaces) and some appropriate to the analytic applications for which a topology is needed [**18**]. We present one particular choice that is useful for topological applications.

Definition 6.13. Suppose $K \subset X$ and $U \subset Y$. Let $S(K, U) = \{f \colon X \to Y$, continuous with $f(K) \subset U\}$. The collection $\mathcal{S} = \{S(K, U) \mid K \subset X$ compact, $U \subset Y$ open$\}$ is a subbasis for the topology $\mathcal{T}_\mathcal{S}$ on $\operatorname{Hom}(X, Y)$ called the **compact-open topology**. We denote the space $(\operatorname{Hom}(X, Y), \mathcal{T}_\mathcal{S})$ as $\operatorname{map}(X, Y)$.

Theorem 6.14. (1) *If X is locally compact and Hausdorff, then the evaluation mapping*

$$e\colon X \times \mathrm{map}(X, Y) \to Y, \qquad e(x, f) = f(x),$$

is continuous.

(2) *If X is locally compact and Hausdorff and Z is another space, then a function $F\colon X \times Z \to Y$ is continuous if and only if its adjoint map $\hat{F}\colon Z \to \mathrm{map}(X, Y)$, defined by $\hat{F}(z)(x) = F(x, z)$, is continuous.*

Proof. Given $(x, f) \in X \times \mathrm{map}(X, Y)$ suppose $f(x) \in V$, an open set in Y. Since $x \in f^{-1}(V)$, use the fact that X is locally compact to find U open in X such that $x \in U \subset \mathrm{cls}\, U \subset f^{-1}(V)$ with $\mathrm{cls}\, U$ compact. Then $(x, f) \in U \times S(\mathrm{cls}\, U, V)$, an open set of $X \times \mathrm{map}(X, Y)$. If $(x_1, f_1) \in U \times S(\mathrm{cls}\, U, V)$, then $f_1(x_1) \in V$ so $e(U \times S(\mathrm{cls}\, U, V)) \subset V$ as needed.

Suppose $\hat{F}$ is continuous. Then F is the composite

$$e \circ (\mathrm{id} \times \hat{F})\colon X \times Z \to X \times \mathrm{map}(X, Y) \to Y,$$

which is continuous.

Suppose F is continuous and consider $\hat{F}\colon Z \to \mathrm{map}(X, Y)$. Let $z \in Z$ and let $S(K, U)$ be a subbasis open set containing $\hat{F}(z)$. We show there is an open set $W \subset Z$, with $z \in W$ and $\hat{F}(W) \subset S(K, U)$. Since $\hat{F}(z) \in S(K, U)$, we have $F(K \times \{z\}) \subset U$. Since F is continuous, it follows that $K \times \{z\} \subset F^{-1}(U)$ and $F^{-1}(U)$ is an open set in $X \times Z$. The subset $K \times \{z\}$ is compact and so the collection of basic open sets contained in $F^{-1}(U) \subset X \times Z$ gives an open cover of $K \times \{z\}$. This cover has a finite subcover $U_1 \times W_1, U_2 \times W_2, \ldots, U_n \times W_n$. Let $W = W_1 \cap W_2 \cap \cdots \cap W_n$ be a nonempty open set in Z since $z \in W_i$ for each i. Furthermore, $K \times W \subset F^{-1}(U)$. If $z' \in W$ and $x \in K$, then $F(x, z') \in U$, and so $\hat{F}(W) \subset S(K, U)$ as desired. $\quad\square$

The description of topology as "rubber-sheet geometry" can be made concrete by picturing $\mathrm{map}(X, Y)$. We want to describe a deformation of one mapping into another. If f and g are in $\mathrm{map}(X, Y)$, then a path in $\mathrm{map}(X, Y)$ joining f and g is a mapping $\lambda\colon [0, 1] \to$

map(X, Y) with $\lambda(0) = f$ and $\lambda(1) = g$. This path encodes the deforming of $f(X)$ to $g(X)$ where at time t the shape is $\lambda(t)(X)$. We can rewrite this path using the adjoint to define an important notion to be developed in later chapters.

Definition 6.15. A **homotopy** between functions $f, g \colon X \to Y$ is a continuous function $H \colon X \times [0, 1] \to Y$ with $H(x, 0) = f(x)$, $H(x, 1) = g(x)$. We say that f **is homotopic to** g if there is a homotopy between them.

Notice that $\hat{H} = \lambda$, a path between f and g in map(X, Y). A homotopy may be thought of as a continuous, one-parameter family of functions deforming f into g.

We record some other important properties of the compact-open topology. The proofs are left to the reader.

Proposition 6.16. *Suppose that X is a locally compact and Hausdorff space.*

(1) If $(\mathrm{Hom}(X, Y), \mathcal{T})$ *is another topology on* $\mathrm{Hom}(X, Y)$ *and the evaluation map*

$$e \colon X \times (\mathrm{Hom}(X, Y), \mathcal{T}) \to Y$$

is continuous, then $\mathcal{T}$ *contains the compact-open topology.*

(2) If Y is locally compact and Hausdorff, then the composition of functions

$$\circ \colon \mathrm{map}(X, Y) \times \mathrm{map}(Y, Z) \to \mathrm{map}(X, Z)$$

is continuous.

(3) If Y is Hausdorff, then the space $\mathrm{map}(X, Y)$ *is Hausdorff.*

Conditions on continuous mappings from X to Y lead to subsets of map(X, Y) that may be endowed with the subspace topology. For example, let map$((X, x_0), (Y, y_0))$ denote the subspace of functions $f \colon X \to Y$ for which $f(x_0) = y_0$. This is the **space of pointed maps**. More generally, if $A \subset X$ and $B \subset Y$, we can define the space of maps of pairs, map$((X, A), (Y, B))$, requiring that $f(A) \subset B$.

Exercises

1. A second countable space is "almost" compact. Prove that when X is second countable, every open cover of X has a countable subcover (Lindelöf's theorem).

2. Show that compact Hausdorff space is *normal* (also labelled T_4), that is, given two disjoint closed subsets of X, say A and B, then there are open sets U and V with $A \subset U$, $B \subset V$, and $U \cap V = \emptyset$.

3. A useful property of compact spaces is the **finite intersection property**. Suppose that $\mathcal{F} = \{F_j \mid j \in J\}$ is a collection of closed subsets of X with the following property: $F_1 \cap \cdots \cap F_k \neq \emptyset$ for every finite subcollection $\{F_1, \ldots, F_k\}$ of $\mathcal{F}$; then $\bigcap_{F \in \mathcal{F}} F \neq \emptyset$. Show that this condition is equivalent to a space being compact. (Hint: Consider the complements of the F_i and the consequence of the intersection being empty.)

4. Suppose X is a compact space and $\{x_1, x_2, x_3, \ldots\}$ is a sequence of points in X. Show that there is a subsequence of $\{x_i\}$ that converges to a point in X.

5. Show the easy direction of Tychonoff's theorem, that is, if $\{X_i \mid i \in J\}$ is a collection of nonempty spaces and the product $\prod_{i \in J} X_i$ is compact, then each X_i is compact.

6. Although the compact subsets of $\mathbb{R}$ are easily determined (closed and bounded), things are very different in $\mathbb{Q} \subset \mathbb{R}$ with the subspace topology. Determine the compact subsets of $\mathbb{Q}$. We can mimic the one-point compactification of $\mathbb{R}$ using $\mathbb{Q}$: Let $\hat{\mathbb{Q}} = \mathbb{Q} \cup \{\infty\}$ be topologized by $T = \{U \subset \mathbb{Q}, U \text{ open, or } \hat{\mathbb{Q}} - K, \text{ where } K \text{ is a compact subset of } \mathbb{Q}\}$. Show that $(\hat{\mathbb{Q}}, T)$ is not Hausdorff. Deduce that $\mathbb{Q}$ is not locally compact.

7. Proposition 6.9 states that if $f\colon X \to Y$ is one-one, onto, and continuous, X is compact, and Y is Hausdorff, then f is a homeomorphism. Show that the condition of Y Hausdorff cannot be relaxed.

8. Prove Proposition 6.16.

Chapter 7

Homotopy and the Fundamental Group

The group G will be called the fundamental group of the manifold V.

HENRI POINCARÉ, 1895

The properties of a topological space that we have developed so far have depended on the choice of topology, the collection of open sets. Taking a different tack, we introduce a different structure, algebraic in nature, associated to a space together with a choice of base point (X, x_0). This structure will allow us to bring to bear the power of algebraic arguments. The fundamental group was introduced by Poincaré in his investigations of the action of a group on a manifold [**66**].

The first step in defining the fundamental group is to study more deeply the relation of homotopy between continuous functions $f_0 \colon X \to Y$ and $f_1 \colon X \to Y$. Recall that f_0 is *homotopic* to f_1, denoted $f_0 \simeq f_1$, if there is a continuous function (a *homotopy*)

$$H \colon X \times [0, 1] \to Y \text{ with } H(x, 0) = f_0(x) \text{ and } H(x, 1) = f_1(x).$$

The choice of notation anticipates an interpretation of the homotopy —if we write $H(x, t) = f_t(x)$, then a homotopy is a deformation of the mapping f_0 into the mapping f_1 through the family of mappings f_t.

Theorem 7.1. *The relation* $f \simeq g$ *is an equivalence relation on the set* $\mathrm{Hom}(X, Y)$ *of continuous mappings from* X *to* Y.

Proof. Let $f \colon X \to Y$ be a given mapping. The homotopy $H(x, t) = f(x)$ is a continuous mapping $H \colon X \times [0, 1] \to Y$ and so $f \simeq f$.

If $f_0 \simeq f_1$ and $H \colon X \times [0, 1] \to Y$ is a homotopy between f_0 and f_1, then the mapping $H' \colon X \times [0, 1] \to Y$ given by $H'(x, t) = H(x, 1 - t)$ is continuous and a homotopy between f_1 and f_0, that is, $f_1 \simeq f_0$.

Finally, for $f_0 \simeq f_1$ and $f_1 \simeq f_2$, suppose that $H_1 \colon X \times [0, 1] \to Y$ is a homotopy between f_0 and f_1, and $H_2 \colon X \times [0, 1] \to Y$ is a homotopy between f_1 and f_2. Define the homotopy $H \colon X \times [0, 1] \to Y$ by

$$H(x, t) = \begin{cases} H_1(x, 2t), & \text{if } 0 \le t \le 1/2, \\ H_2(x, 2t - 1), & \text{if } 1/2 \le t \le 1. \end{cases}$$

Since $H_1(x, 1) = f_1(x) = H_2(x, 0)$, the piecewise definition of H gives a continuous function (Theorem 4.4). By definition, $H(x, 0) = f_0(x)$ and $H(x, 1) = f_2(x)$ and so $f_0 \simeq f_2$. $\qquad \square$

We denote the equivalence class under homotopy of a mapping $f \colon X \to Y$ by $[f]$ and the *set of homotopy classes of maps between* X *and* Y by $[X, Y]$. If $F \colon W \to X$ and $G \colon Y \to Z$ are continuous mappings, then the sets $[X, Y]$, $[W, X]$, and $[Y, Z]$ are related.

Proposition 7.2. *Continuous mappings* $F \colon W \to X$ *and* $G \colon Y \to Z$ *induce well-defined functions* $F^* \colon [X, Y] \to [W, Y]$ *and* $G_* \colon [X, Y] \to [X, Z]$ *by* $F^*([h]) = [h \circ F]$ *and* $G_*([h]) = [G \circ h]$ *for* $[h] \in [X, Y]$.

Proof. We need to show that if $h \simeq h'$, then $h \circ F \simeq h' \circ F$ and $G \circ h \simeq G \circ h'$. Fixing a homotopy $H \colon X \times [0, 1] \to Y$ with $H(x, 0) = h(x)$ and $H(x, 1) = h'(x)$, then the desired homotopies are $H_F(w, t) = H(F(w), t)$ and $H_G(x, t) = G(H(x, t))$. $\qquad \square$

To a space X we associate a space particularly rich in structure, the *mapping space of paths in* X, $\mathrm{map}([0, 1], X)$. Recall that $\mathrm{map}([0, 1], X)$ is the set of continuous mappings $\mathrm{Hom}([0, 1], X)$ with the compact-open topology. The space $\mathrm{map}([0,1], X)$ has the following properties.

(1) X embeds into $\mathrm{map}([0,1], X)$ by associating to each point $x \in X$ the *constant path* $c_x(t) = x$ for all $t \in [0,1]$.

(2) Given a path $\lambda \colon [0,1] \to X$, we can *reverse* the path by composing with $t \mapsto 1 - t$. Let $\lambda^{-1}(t) = \lambda(1-t)$.

(3) Given a pair of paths λ, $\mu \colon [0,1] \to X$ for which $\lambda(1) = \mu(0)$, we can *compose* paths by

$$
\lambda * \mu(t) = \begin{cases} \lambda(2t), & \text{if } 0 \leq t \leq 1/2, \\ \mu(2t - 1), & \text{if } 1/2 \leq t \leq 1. \end{cases}
$$

Thus, for certain pairs of paths λ and μ, we obtain a new path $\lambda * \mu \in \mathrm{map}([0,1], X)$.

Composition of paths is always defined when we restrict to a certain subspace of $\mathrm{map}([0,1], X)$.

Definition 7.3. Suppose X is a space and $x_0 \in X$ is a choice of base point in X. The **space of based loops** in X, denoted $\Omega(X, x_0)$, is the subspace of $\mathrm{map}([0,1], X)$,

$$
\Omega(X, x_0) = \{ \lambda \in \mathrm{map}([0,1], X) \mid \lambda(0) = \lambda(1) = x_0 \}.
$$

Composition of loops determines a binary operation $* \colon \Omega(X, x_0) \times \Omega(X, x_0) \to \Omega(X, x_0)$.

We restrict the notion of homotopy when applied to the space of based loops in X in order to stay in that space during the deformation.

Definition 7.4. Given two based loops λ and μ, a **loop homotopy** between them is a homotopy of paths $H \colon [0,1] \times [0,1] \to X$ with $H(t,0) = \lambda(t)$, $H(t,1) = \mu(t)$, and $H(0,s) = H(1,s) = x_0$. That is, for each $s \in [0,1]$, the path $t \mapsto H(t,s)$ is a loop at x_0.

The relation of loop homotopy on $\Omega(X, x_0)$ is an equivalence relation; the proof follows the proof of Theorem 7.1. We denote the set of equivalence classes under loop homotopy by $\pi_1(X, x_0) = [\Omega(X, x_0)]$, a notation for the first of a family of such sets, to be explained later. As it turns out, $\pi_1(X, x_0)$ enjoys some remarkable properties.

Theorem 7.5. *Composition of loops induces a group structure on* $\pi_1(X, x_0)$ *with identity element* $[c_{x_0}(t)]$ *and inverses given by* $[\lambda]^{-1} = [\lambda^{-1}]$.

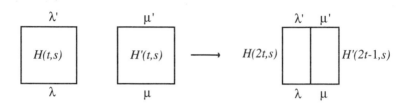

Proof. We begin by showing that composition of loops induces a well-defined binary operation on the homotopy classes of loops. Given $[\lambda]$ and $[\mu]$, we then define $[\lambda] * [\mu] = [\lambda * \mu]$. Suppose that $[\lambda] = [\lambda']$ and $[\mu] = [\mu']$. We must show that $\lambda * \mu \simeq \lambda' * \mu'$. If $H \colon [0,1] \times [0,1] \to X$ is a loop homotopy between λ and λ' and $H' \colon [0,1] \times [0,1] \to X$ a loop homotopy between μ and μ', then form $H'' \colon [0,1] \times [0,1] \to X$ defined by

$$H''(t, s) = \begin{cases} H(2t, s), & \text{if } 0 \leq t \leq 1/2, \\ H'(2t - 1, s), & \text{if } 1/2 \leq t \leq 1. \end{cases}$$

Since $H''(0, s) = H(0, s) = x_0$ and $H''(1, s) = H'(1, s) = x_0$, H'' is a loop homotopy. Also $H''(t, 0) = \lambda * \mu(t)$ and $H''(t, 1) = \lambda' * \mu'(t)$, and the binary operation is well defined on equivalence classes of loops.

We next show that $*$ is associative. Notice that $(\lambda * \mu) * \nu \neq \lambda * (\mu * \nu)$; we only get $1/4$ of the interval for λ in the first product and $1/2$ of the interval in the second product. We define the explicit homotopy after its picture, which makes the point more clearly:

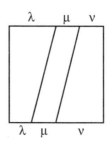

$$H(t,s) = \begin{cases} \lambda(4t/(1+s)), & \text{if } 0 \le t \le (s+1)/4, \\ \mu(4t-1-s), & \text{if } (s+1)/4 \le t \le (s+2)/4, \\ \nu\left(1 - \dfrac{4(1-t)}{(2-s)}\right), & \text{if } (s+2)/4 \le t \le 1. \end{cases}$$

The class of the constant map, $e(t) = c_{x_0}(t) = x_0$, gives the identity for $\pi_1(X, x_0)$. To see this, we show, for all $\lambda \in \Omega(X, x_0)$, that $\lambda * e \simeq \lambda \simeq e * \lambda$ via loop homotopies. This is accomplished in the case $\lambda \simeq e * \lambda$ by the homotopy:

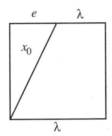

$$F(t,s) = \begin{cases} x_0, & \text{if } 0 \le t \le s/2, \\ \lambda((2t-s)/(2-s)), & \text{if } s/2 \le t \le 1. \end{cases}$$

The case $\lambda \simeq \lambda * e$ is similar. Finally, inverses are constructed by using the reverse loop $\lambda^{-1}(t) = \lambda(1-t)$. To show that $\lambda * \lambda^{-1} \simeq e$ consider the homotopy:

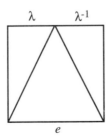

$$G(t,s) = \begin{cases} \lambda(2t), & \text{if } 0 \le t \le s/2, \\ \lambda(s), & \text{if } s/2 \le t \le 1 - (s/2), \\ \lambda(2-2t), & \text{if } 1-(s/2) \le t \le 1. \end{cases}$$

The homotopy resembles the loop, moving out for a while, waiting a little, and then shrinking back along itself. The proof that $\lambda^{-1} * \lambda \simeq e$ is similar. $\square$

Definition 7.6. The group $\pi_1(X, x_0)$ is called the **fundamental group of X at the base point** x_0.

Suppose x_1 is another choice of basepoint for X. If X is path-connected, there is a path $\gamma \colon [0, 1] \to X$ with $\gamma(0) = x_0$ and $\gamma(1) = x_1$. This path induces a mapping $u_\gamma \colon \pi_1(X, x_0) \to \pi_1(X, x_1)$ by $[\lambda] \mapsto [\gamma^{-1} * \lambda * \gamma]$, that is, follow γ^{-1} from x_1 to x_0, then follow λ around and back to x_0, then follow γ back to x_1, all giving a loop based at x_1. Notice

$$
\begin{aligned}
u_\gamma([\lambda] * [\mu]) &= u_\gamma([\lambda * \mu]) \\
&= [\gamma^{-1} * \lambda * \mu * \gamma] \\
&= [\gamma^{-1} * \lambda * \gamma * \gamma^{-1} * \mu * \gamma] \\
&= [\gamma^{-1} * \lambda * \gamma] * [\gamma^{-1} * \mu * \gamma] = u_\gamma([\lambda]) * u_\gamma([\mu]).
\end{aligned}
$$

Thus u_γ is a homomorphism. The mapping $u_{\gamma^{-1}} \colon \pi_1(X, x_1) \to \pi_1(X, x_0)$ is an inverse, since $[\gamma * (\gamma^{-1} * \lambda * \gamma) * \gamma^{-1}] = [\lambda]$. Thus $\pi_1(X, x_0)$ is isomorphic to $\pi_1(X, x_1)$ whenever x_0 is joined to x_1 by a path. Though it is a bit of a lie, we write $\pi_1(X)$ for a space X that is path-connected since any choice of basepoint gives an isomorphic group. In this case, $\pi_1(X)$ denotes an isomorphism class of groups.

Following Proposition 7.2, a continuous function $f \colon X \to Y$ induces a mapping

$$
f_* \colon \pi_1(X, x_0) \to \pi_1(Y, f(x_0)), \text{ given by } f_*([\lambda]) = [f \circ \lambda].
$$

In fact, f_* is a homomorphism of groups:

$$
\begin{aligned}
f_*([\lambda] * [\mu]) &= f_*([\lambda * \mu]) = [f \circ (\lambda * \mu)] \\
&= [(f \circ \lambda) * (f \circ \mu)] = [f \circ \lambda] * [f \circ \mu] \\
&= f_*([\lambda]) * f_*([\mu]).
\end{aligned}
$$

Furthermore, when we have continuous mappings $f \colon X \to Y$ and $g \colon Y \to Z$, we obtain $f_* \colon \pi_1(X, x_0) \to \pi_1(Y, f(x_0))$ and $g_* \colon \pi_1(Y, f(x_0)) \to \pi_1(Z, g \circ f(x_0))$. Observe that

$$
g_* \circ f_*([\lambda]) = g_*([f \circ \lambda]) = [g \circ f \circ \lambda] = (g \circ f)_*([\lambda]),
$$

so we have $(g \circ f)_* = g_* \circ f_*$. It is evident that the identity mapping id: $X \to X$ induces the identity homomorphism of groups $\pi_1(X, x_0) \to \pi_1(X, x_0)$. We can summarize these observations by the (post-1945) remark that π_1 *is a functor from pointed spaces and pointed maps to groups and group homomorphisms.* Since we are focusing on classical notions in topology (pre-1935) and category theory was christened later, we will not use this language in what follows. For an introduction to this framework see [**53**].

The behavior of the induced homomorphisms under composition has the following consequence.

Corollary 7.7. *The fundamental group is a topological invariant of a space. That is, if $f: X \to Y$ is a homeomorphism, then the groups $\pi_1(X, x_0)$ and $\pi_1(Y, f(x_0))$ are isomorphic.*

Proof. Suppose $f: X \to Y$ has continuous inverse $g: Y \to X$. Then $g \circ f = \mathrm{id}_X$ and $f \circ g = \mathrm{id}_Y$. It follows that $g_* \circ f_* = \mathrm{id}$ and $f_* \circ g_* = \mathrm{id}$ on $\pi_1(X, x_0)$ and $\pi_1(Y, f(x_0))$, respectively. Thus f_* and g_* are group isomorphisms. $\qquad\square$

Before we do some calculations we derive a few more formal properties of the fundamental group. In particular, what conditions imply $\pi_1(X) = \{e\}$, and how does the fundamental group behave under the formation of subspaces, products, and quotients?

Definition 7.8. A subspace $A \subset X$ is a **retract** of X if there is a continuous function, the retraction, $r: X \to A$ for which $r(a) = a$ for all $a \in A$. The subset $A \subset X$ is a **deformation retraction** if A is a retract of X and the composition $i \circ r: X \to A \hookrightarrow X$ is homotopic to the identity on X via a homotopy that fixes A, that is, there is a homotopy $H: X \times [0,1] \to X$ with

$$H(x, 0) = x, H(x, 1) = r(x), \text{ and } H(a, t) = a$$

for all $a \in A$ and all $t \in [0, 1]$.

Proposition 7.9. *If $A \subset X$ is a retract with retraction $r: X \to A$, then the inclusion $i: A \to X$ induces an injective homomorphism $i_*: \pi_1(A, a) \to \pi_1(X, a)$ and the retraction induces a surjective homomorphism $r_*: \pi_1(X, a) \to \pi_1(A, a)$.*

Proof. The composite $r \circ i \colon A \to X \to A$ is the identity mapping on A and so the composite $r_* \circ i_* \colon \pi_1(A, a) \to \pi_1(X, a) \to \pi_1(A, a)$ is the identity on $\pi_1(A, a)$. If $i_*([\lambda]) = i_*([\lambda'])$, then $[\lambda] = r_* i_*([\lambda]) = r_* i_*([\lambda']) = [\lambda']$, and so the homomorphism i_* is injective. If $[\lambda] \in \pi_1(A, a)$, then $r_*(i_*([\lambda])) = [\lambda]$ and so r_* is onto. □

Examples. Represent the Möbius band M by glueing the left and right edges of $[0, 1] \times [0, 1]$ with a twist (Chapter 4). Let $A = \{[(t, \frac{1}{2})] \mid 0 \le t \le 1\} \subset M$ be the circle in the middle of the band. After the identification, A is homeomorphic to S^1. Define the map $r \colon M \to A$ by projecting straight down or up to this line, that is, $[(t, s)] \mapsto [(t, \frac{1}{2})]$. It is easy to see that r is continuous and $r|_A = \mathrm{id}_A$ so we have a retract. Thus the composite $r_* \circ i_* \colon \pi_1(S^1) \to \pi_1(M) \to \pi_1(S^1)$ is the identity on $\pi(S^1)$.

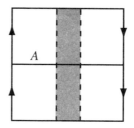

For any space X, the inclusion followed by projection
$$X \cong X \times \{0\} \hookrightarrow X \times [0, 1] \to X$$
is the identity and so X is a retract of $X \times [0, 1]$. In fact, X is a deformation retraction via the deformation $H \colon X \times [0, 1] \times [0, 1] \to X \times [0, 1]$ given by $H(x, t, s) = (x, ts)$: when $s = 1$, $H(x, t, 1) = (x, t)$ and for $s = 0$ we have $H(x, t, 0) = (x, 0)$.

Recall that a subset K of $\mathbb{R}^n$ is **convex** if whenever $\mathbf{x}$ and $\mathbf{y}$ are in K, then for all $t \in [0, 1]$, $t\mathbf{x} + (1 - t)\mathbf{y} \in K$. If $K \subset \mathbb{R}^n$ is convex, let $\mathbf{x}_0 \in K$. Then K is a deformation retraction of the one-point subset $\{\mathbf{x}_0\}$ by the homotopy $H(\mathbf{x}, t) = t\mathbf{x}_0 + (1 - t)\mathbf{x}$. When $t = 0$ we have $H(\mathbf{x}, 0) = \mathbf{x}$ and when $t = 1$, $H(\mathbf{x}, 1) = \mathbf{x}_0$. The retraction $K \to \{\mathbf{x}_0\}$ is thus a deformation of the identity on K. Examples of convex subsets of $\mathbb{R}^n$ include $\mathbb{R}^n$ itself, any open ball $B(\mathbf{x}, \epsilon)$, and the boxes $[a_1, b_1] \times \cdots \times [a_n, b_n]$.

More generally, there is always the retract $\{x_0\} \hookrightarrow X \to \{x_0\}$, which leads to the trivial homomorphisms of groups $\{e\} \to \pi_1(X, x_0) \to \{e\}$. This retract is not always a deformation retract. We call a space **contractible** when it is a deformation retract of one of its points.

Deformation retracts give isomorphic fundamental groups.

Theorem 7.10. *If A is a deformation retract of X, then the inclusion $i: A \to X$ induces an isomorphism $i_*: \pi_1(A, a) \to \pi_1(X, a)$.*

Proof. From the definition of a deformation retract, the composite $i \circ r: X \to A \hookrightarrow X$ is homotopic to id_X via a homotopy fixing the points in A, that is, there is a homotopy $H: X \times [0, 1] \to X$ with $H(x, 0) = i \circ r(x)$, $H(x, 1) = x$, and $H(a, t) = a$ for all $t \in [0, 1]$. We show that $i_* \circ r_*([\lambda]) = [\lambda]$. In fact we show a little more:

Lemma 7.11. *If $f, g: (X, x_0) \to (Y, y_0)$ are continuous functions, homotopic through basepoint preserving maps, then $f_* = g_*: \pi_1(X, x_0) \to \pi_1(Y, y_0)$.*

Proof. Suppose there is a homotopy $G: X \times [0, 1] \to Y$ with $G(x, 0) = f(x)$, $G(x, 1) = g(x)$, and $G(x_0, t) = y_0$ for all $t \in [0, 1]$. Consider a loop based at x_0, $\lambda: [0, 1] \to X$, and the compositions $f \circ \lambda$, $g \circ \lambda$, and $G \circ (\lambda \times \mathrm{id}): [0, 1] \times [0, 1] \to Y$:

$$G(\lambda(s), 0) = f \circ \lambda(s),$$
$$G(\lambda(s), 1) = g \circ \lambda(s),$$
$$G(\lambda(0), t) = G(\lambda(1), t) = y_0 \text{ for all } t \in [0, 1].$$

Thus $f_*[\lambda] = [f \circ \lambda] = [g \circ \lambda] = g_*[\lambda]$. Hence $f_* = g_*: \pi_1(X, x_0) \to \pi_1(Y, y_0)$. $\square$

A deformation retract gives a basepoint preserving homotopy between $i \circ r$ and id_X, so we have $\mathrm{id} = i_* \circ r_*: \pi_1(X, a) \to \pi_1(X, a)$. By Proposition 7.9, we already know i_* is injective; i_* is surjective because for $[\lambda]$ any class in $\pi_1(X, a)$, one has $[\lambda] = i_*(r_*([\lambda]))$. $\square$

Examples. A convex subset of $\mathbb{R}^n$ is a deformation retract of any point $\mathbf{x}_0$ in the set. It follows from $\pi_1(\{\mathbf{x}_0\}) = \{e\}$, that for any convex subset $K \subset \mathbb{R}^n$, $\pi_1(K, \mathbf{x}_0) = \{e\}$. Of course, this includes

$\pi_1(\mathbb{R}^n, \mathbf{0}) = \{e\}$. Next consider $\mathbb{R}^n - \{\mathbf{0}\}$. The $(n-1)$-sphere $S^{n-1} \subset \mathbb{R}^n$ is a deformation retract of $\mathbb{R}^n - \{\mathbf{0}\}$ as follows: Let

$$F \colon (\mathbb{R}^n - \{\mathbf{0}\}) \times [0,1] \to \mathbb{R}^n - \{\mathbf{0}\}$$

be given by

$$F(\mathbf{x}, t) = (1-t)\mathbf{x} + t \frac{\mathbf{x}}{\|\mathbf{x}\|}.$$

Here $F(\mathbf{x}, 0) = \mathbf{x}$ and $F(\mathbf{x}, 1) = \mathbf{x}/\|\mathbf{x}\| \in S^{n-1}$. By Theorem 7.10,

$$\pi_1(\mathbb{R}^n - \{\mathbf{0}\}, \mathbf{x}_0/\|\mathbf{x}_0\|) \cong \pi_1(S^{n-1}, \mathbf{x}_0/\|\mathbf{x}_0\|).$$

A space X is said to be **simply-connected** (or *1-connected*) if it is path-connected and $\pi_1(X) = \{e\}$. Any convex subset of $\mathbb{R}^n$ or, more generally, any contractible space is simply-connected. Furthermore, simple connectivity is a topological property.

Theorem 7.12. *Suppose $X = U \cup V$, where U and V are open, simply-connected subspaces and $U \cap V$ is path-connected. Then X is simply-connected.*

Proof. Choose a point $x_0 \in U \cap V$ as basepoint. Let $\lambda \colon [0,1] \to X$ be a loop based at x_0. Since λ is continuous, $\{\lambda^{-1}(U), \lambda^{-1}(V)\}$ is an open cover of the compact space $[0,1]$. The Lebesgue Lemma gives points $0 = t_0 < t_1 < t_2 < \cdots < t_n = 1$ with $\lambda([t_{i-1}, t_i]) \subset U$ or V. We can join x_0 to $\lambda(t_i)$ by a path γ_i. Define for $i \geq 1$,

$$\lambda_i(s) = \lambda((t_i - t_{i-1})s + t_{i-1}), \quad 0 \leq s \leq 1,$$

for the path along λ joining $\lambda(t_{i-1})$ to $\lambda(t_i)$.

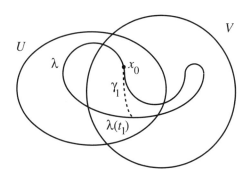

Then $\lambda \simeq \lambda_1 * \lambda_2 * \cdots * \lambda_n$ and, furthermore,

$$\lambda \simeq (\lambda_1 * \gamma_1^{-1}) * (\gamma_1 * \lambda_2 * \gamma_2^{-1}) * (\gamma_2 * \lambda_3 * \gamma_3^{-1}) * \cdots * (\gamma_{n-1} * \lambda_n).$$

Each $\gamma_i * \lambda_{i+1} * \gamma_{i+1}^{-1}$ lies in U or V. Since U and V are simply-connected, each of these loops is homotopic to the constant map. Thus $\lambda \simeq c_{x_0}$. It follows that $\pi_1(X, x_0) \cong \{e\}$. $\qquad\square$

Corollary 7.13. $\pi_1(S^n) \cong \{e\}$ *for* $n \geq 2$.

Proof. We can decompose S^n as a union of $U = \{(r_0, r_1, \ldots, r_n) \in S^n \mid r_n > -1/4\}$ and $V = \{(r_0, r_1, \ldots, r_n) \in S^n \mid r_n < 1/4\}$. By stereographic projection from each pole, we can establish that U and V are homeomorphic to an open disk in $\mathbb{R}^n$, which is convex. The intersection $U \cap V$ is homeomorphic to $S^{n-1} \times (-1/4, 1/4)$, which is path-connected when $n \geq 2$. $\qquad\square$

Since $S^{n-1} \subset \mathbb{R}^n - \{\mathbf{0}\}$ is a deformation retract, we have proven:

Corollary 7.14. $\pi_1(\mathbb{R}^n - \{\mathbf{0}\}) \cong \{e\}$ *for* $n \geq 3$.

In Chapter 8 we will consider the case $\pi_1(S^1)$ in detail.

We next consider the fundamental group of a product $X \times Y$.

Theorem 7.15. *Let* (X, x_0) *and* (Y, y_0) *be pointed spaces. Then* $\pi_1(X \times Y, (x_0, y_0))$ *is isomorphic to* $\pi_1(X, x_0) \times \pi_1(Y, y_0)$, *the direct product of these two groups.*

Recall that if G and H are groups, the direct product $G \times H$ has underlying set the cartesian product of G and H and binary operation $(g_1, h_1) \cdot (g_2, h_2) = (g_1 g_2, h_1 h_2)$.

Proof. Recall from Chapter 4 that a mapping $\lambda \colon [0, 1] \to X \times Y$ is continuous if and only if $\mathrm{pr}_1 \circ \lambda \colon [0, 1] \to X$ and $\mathrm{pr}_2 \circ \lambda \colon [0, 1] \to Y$ are continuous. If λ is a loop at (x_0, y_0), then $\mathrm{pr}_1 \circ \lambda$ is a loop at x_0 and $\mathrm{pr}_2 \circ \lambda$ is a loop at y_0. We leave it to the reader to prove that

1) If $\lambda \simeq \lambda' \colon [0, 1] \to X \times Y$, then $\mathrm{pr}_i \circ \lambda \simeq \mathrm{pr}_i \circ \lambda'$ for $i = 1, 2$.

2) If we take $\lambda * \lambda' \colon [0, 1] \to X \times Y$, then $\mathrm{pr}_i \circ (\lambda * \lambda') = (\mathrm{pr}_i \circ \lambda) * (\mathrm{pr}_i \circ \lambda')$.

These facts allow us to define a homomorphism

$$\mathrm{pr}_{1*} \times \mathrm{pr}_{2*} \colon \pi_1(X \times Y, (x_0, y_0)) \to \pi_1(X, x_0) \times \pi_1(Y, y_0)$$

by $\mathrm{pr}_{1*} \times \mathrm{pr}_{2*}([\lambda]) = ([\mathrm{pr}_1 \circ \lambda], [\mathrm{pr}_2 \circ \lambda])$. The inverse homomorphism is given by $([\lambda], [\mu]) \mapsto [(\lambda, \mu)(t)]$, where $(\lambda, \mu)(t) = (\lambda(t), \mu(t))$. Thus we have an isomorphism. $\qquad\square$

We can use such results to show that certain subspaces of a space are *not* deformation retracts. For example, if $\pi_1(X, x_0)$ is a nontrivial group, then $\pi_1(X \times X, (x_0, x_0))$ is not isomorphic to $\pi_1(X \times \{x_0\}, (x_0, x_0))$. Although $X \times \{x_0\}$ is a retract of $X \times X$ via

$$X \times \{x_0\} \hookrightarrow X \times X \to X \times \{x_0\},$$

it is not a deformation retract of $X \times X$.

Extra structure on a space can lead to more structure on the fundamental group. Recall (exercises of Chapter 4) that a topological group, (G, e), is a Hausdorff topological space with basepoint $e \in G$ together with a continuous function (the group operation) $m \colon G \times G \to G$, satisfying $m(g, e) = m(e, g) = g$ for all $g \in G$, as well as another continuous function (the inverse) $\mathrm{inv} \colon G \to G$ with $m(g, \mathrm{inv}(g)) = e = m(\mathrm{inv}(g), g)$ for all $g \in G$.

Theorem 7.15 allows us to define a new binary operation on $\pi_1(G, e)$, the composite of the isomorphism of the theorem with the homomorphism induced by m:

$$\mu_* \colon \pi_1(G, e) \times \pi_1(G, e) \to \pi_1(G \times G, (e, e)) \to \pi_1(G, e).$$

We denote the binary operation by $\mu_*([\lambda], [\nu]) = [\lambda \natural \nu]$. On the level of loops, this mapping is given explicitly by $(\lambda, \mu) \mapsto \lambda \natural \mu$, where $(\lambda \natural \mu)(t) = m(\lambda(t), \mu(t))$. We next compare this binary operation with the usual multiplication of loops for the fundamental group.

Theorem 7.16. *If G is a topological group, then $\pi_1(G, e)$ is an abelian group.*

Proof. We first show that $\sharp$ and the usual multiplication $*$ on $\pi_1(G,e)$ are actually the same binary operation! We argue as follows: Represent $\lambda * \mu(t)$ by $\lambda' \sharp \mu'(t)$, where

$$\lambda'(t) = \begin{cases} \lambda(2t), & 0 \le t \le \frac{1}{2}, \\ e, & \frac{1}{2} \le t \le 1, \end{cases} \qquad \mu'(t) = \begin{cases} e, & 0 \le t \le \frac{1}{2}, \\ \mu(2t-1), & \frac{1}{2} \le t \le 1. \end{cases}$$

Since $\lambda(1) = e = \mu(0)$ and $m(e, \mu'(t)) = \mu'(t)$, $m(\lambda'(t), e) = \lambda'(t)$, we see $\lambda * \mu(t) = m(\lambda'(t), \mu'(t))$. We next show that $\lambda * \mu$ is loop homotopic to $\lambda \sharp \mu$. Define two functions $h_1, h_2 \colon [0,1] \times [0,1] \to [0,1]$ by

$$h_1(t,s) = \begin{cases} 2t/(2-s), & 0 \le t \le 1 - (s/2), \\ 1, & 1 - s/2 \le t \le 1, \end{cases}$$

$$h_2(t,s) = \begin{cases} 0, & 0 \le t \le s/2, \\ (2t-s)/(2-s), & s/2 \le t \le 1. \end{cases}$$

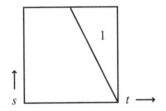
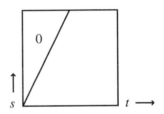

Let $F(t,s) = m(\lambda(h_1(t,s)), \mu(h_2(t,s)))$. Since it is a composition of continuous functions, F is continuous. Notice

$$F(t,0) = m(\lambda(h_1(t,0)), \mu(h_2(t,0))) = m(\lambda(t), \mu(t)) = \lambda \sharp \mu(t)$$

and

$$F(t,1) = m(\lambda(h_1(t,1)), \mu(h_2(t,1))) = m(\lambda'(t), \mu'(t)) = \lambda * \mu(t).$$

Thus $\lambda * \mu$ is loop homotopic to $\lambda \sharp \mu$ and we get the same binary operation.

Given two loops λ and μ, consider the function

$$\mathcal{G} \colon [0,1] \times [0,1] \to G, \quad \mathcal{G}(t,s) = m(\lambda(t), \mu(s)).$$

The four corners are mapped to e and the diagonal from the lower left to the upper right is given by $\lambda \sharp \mu$. We will take some liberties

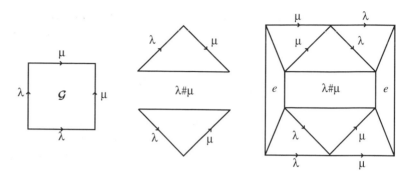

and argue with diagrams to construct a loop homotopy from $\lambda * \mu$ to $\mu * \lambda$.

Slice the square filled in by $\mathcal{G}$ along the diagonal and paste in a rectangle that is simply a product of $\lambda \,\natural\, \mu$ with an interval. Put the resulting hexagon into a square and fill in the remaining regions as the constant map at e, the identity element of G, in the trapezoidal regions and as λ or μ in the triangles where the path lies along the lines joining a vertex to the opposite side.

The diagram gives a homotopy from $\lambda * \mu$ to $\mu * \lambda$. It follows then that $[\lambda] * [\mu] = [\mu] * [\lambda]$ and so $\pi_1(G, e)$ is abelian.   □

Since S^1 is the topological group of unit length complex numbers, we have proved:

Corollary 7.17. $\pi_1(S^1, 1)$ *is abelian.*

Exercises

1. The unit sphere in $\mathbb{R}$ is the set $S^0 = \{-1, 1\}$. Show that the set of homotopy classes of basepoint preserving mappings $[(S^0, -1), (X, x_0)]$ is the same set as $\pi_0(X)$, the set of path components of X.

2. Suppose that $f \colon X \to S^2$ is a continuous mapping that is *not* onto. Show that f is homotopic to a constant mapping.

3. If X is a space, recall that the *cone on* X is the quotient space $CX = X \times [0, 1]/X \times \{1\}$. Suppose $f \colon X \to Y$ is a continuous function and f is homotopic to a constant mapping

$c_y \colon X \to Y$ for some $y \in Y$. Show that there is an extension of f, $\hat{f} \colon CX \to Y$, so that $f = \hat{f} \circ i$, where $i \colon X \to CX$ is the inclusion, $i(x) = [(x, 0)]$.

4. Suppose that X is a path-connected space. When is it true that for any pair of points $p, q \in X$ all paths from p to q induce the same isomorphism between $\pi_1(X, p)$ and $\pi_1(X, q)$?

5. Prove that a disk minus two points is a deformation retract of a figure 8 (that is, $S^1 \vee S^1$).

6. A *starlike space* is a slightly weaker notion than a convex space—in a starlike space $X \subset \mathbb{R}^n$, there is a point $x_0 \in X$ so that for any other point $y \in X$ and any $t \in [0, 1]$ the point $tx_0 + (1 - t)y$ is in X. Give an example of a starlike space that is not convex. Show that a starlike space is a deformation retract of a point.

7. If $K = \alpha(S^1) \subset \mathbb{R}^3$ is a knot, that is, a homeomorphic image of a circle in $\mathbb{R}^3$, then the complement of the knot $\mathbb{R}^3 - K$ has fundamental group $\pi_1(\mathbb{R}^3 - K)$. In fact, this group is an invariant of the knot in a sense that can be made precise. Give a plausibility argument that $\pi_1(\mathbb{R}^2 - K) \neq \{0\}$. See [**69**] for a thorough treatment of this important invariant of knots.

Chapter 8

Computations and Covering Spaces

> *... it is necessary, in order to affirm that a manifold is simply-connected, to study its* fundamental group, *...*

HENRI POINCARÉ, 1904

We have defined the fundamental group and showed that it is a topological invariant, that is, homeomorphic spaces have isomorphic fundamental groups. But we have yet to consider a space whose fundamental group is nontrivial. Two familiar spaces, S^1 and $\mathbb{R}P^2$, will provide examples.

The method of computation focuses on the properties of the mappings,

$$w\colon \mathbb{R} \to S^1 \qquad \qquad p\colon S^2 \to \mathbb{R}P^2$$
$$w(r) = \cos(2\pi r) + i\sin(2\pi r) = e^{2\pi i r} \quad \text{and} \quad p(\mathbf{x}) = [\pm\mathbf{x}].$$

These mappings share certain important properties.

Definition 8.1. Let X be a space. A **covering space** of X is a path-connected space $\tilde{X}$ and a mapping $p\colon \tilde{X} \to X$ such that, for every $x \in X$, there is an open, path-connected subset U with $x \in U$ for which each path component of $p^{-1}(U)$ is homeomorphic to U by restriction of the mapping p. Such open sets are called **elementary neighborhoods**.

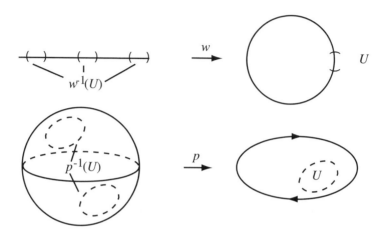

For example, if $e^{i\theta} \in S^1$, then for $0 < \epsilon < \pi$, the open set $U = \{e^{i\alpha} \mid \theta - \epsilon < \alpha < \theta + \epsilon\}$ in S^1 has inverse image under w given by

$$w^{-1}(U) = \bigcup_{k \in \mathbb{Z}} \left(\frac{\theta}{2\pi} - \frac{\epsilon}{2\pi} + k, \frac{\theta}{2\pi} + \frac{\epsilon}{2\pi} + k \right).$$

Since $\epsilon/2\pi < 1/2$, the intervals in the union are all disjoint. Furthermore, w restricted to any one of these intervals has an inverse given by a branch of the logarithm. In the case of the quotient map $p \colon S^2 \to \mathbb{RP}^2$, for a connected open set $V \subset S^2$ satisfying $V \cap -V = \emptyset$, we have $p(V)$ open in $\mathbb{RP}^2$ and $p^{-1}(p(V)) = V \cup -V$. Since the components of $p^{-1}(p(V))$ are V and $-V$, it is an elementary neighborhood. For any $[\pm \mathbf{x}] \in \mathbb{RP}^2$, there is such an elementary neighborhood containing $[\pm \mathbf{x}]$ and so $p \colon S^2 \to \mathbb{RP}^2$ is a covering space.

Henceforth we will assume that all spaces are path-connected and locally path-connected to avoid pathological cases. The most useful property of covering spaces is the ability to lift paths in X to paths in $\tilde{X}$ while preserving the homotopy relation.

Lemma 8.2. *Let $p \colon \tilde{X} \to X$ be a covering space and let $\tilde{x}_0 \in \tilde{X}$ with $p(\tilde{x}_0) = x_0 \in X$. If $\lambda \colon [0,1] \to X$ is any path with $\lambda(0) = x_0$, then there exists a unique path $\hat{\lambda} \colon [0,1] \to \tilde{X}$ with $\hat{\lambda}(0) = \tilde{x}_0$ and $p \circ \hat{\lambda} = \lambda$.*

Proof. Cover X by elementary neighborhoods. If $\lambda([0,1]) \subset U$ for some elementary neighborhood, then $x_0 \in U$ and $\tilde{x}_0 \in p^{-1}(U)$. It follows that $\tilde{x}_0$ lies in some component C_0 of $p^{-1}(U)$ that is homeomorphic to U via $p|_{C_0} : C_0 \to U$. Let $(p|_{C_0})^{-1} : U \to C_0$ denote the inverse of this homeomorphism and let $\hat{\lambda} = (p|_{C_0})^{-1} \circ \lambda$. Then $\hat{\lambda}(0) = (p|_{C_0})^{-1}(x_0) = \tilde{x}_0$, since $\tilde{x}_0$ is the only point in $\tilde{X}$ corresponding to x_0 in this component. Finally, $p \circ \hat{\lambda} = p \circ (p|_{C_0})^{-1} \circ \lambda = \lambda$.

If $\lambda([0,1]) \not\subset U$, consider the collection $\{\lambda^{-1}(U') \subset [0,1] \mid U',$ an elementary neighborhood$\}$. This is a cover of $[0,1]$, which is a compact metric space, and so by Lebesgue's Lemma we can choose $0 = t_0 < t_1 < \cdots < t_{n-1} < t_n = 1$ with each $\lambda([t_{i-1}, t_i])$ a subset of some elementary neighborhood (take $t_i - t_{i-1} < \delta$, the Lebesgue number). Using the argument above, lift λ on $[0, t_1]$. Then take $\lambda(t_1)$ as x_0 and $\hat{\lambda}(t_1)$ as $\tilde{x}_0$ and lift λ to $[t_1, t_2]$. Continuing in this manner, we construct $\hat{\lambda}$ on $[0,1]$ with $\hat{\lambda}(0) = \tilde{x}_0$ and $p \circ \hat{\lambda} = \lambda$.

To show that $\hat{\lambda}$ constructed in this manner is unique, we prove a more general result that implies uniqueness.

Lemma 8.3. *Let* $p \colon \tilde{X} \longrightarrow X$ *be a covering space and* Y *a connected, locally connected space. Given two mappings* f_1, $f_2 \colon Y \to \tilde{X}$ *with* $p \circ f_1 = p \circ f_2$, *then the set* $\{y \in Y \mid f_1(y) = f_2(y)\}$ *is either empty or all of* Y.

Proof. Consider the subset of Y given by $B = \{y \in Y \mid f_1(y) = f_2(y)\}$. We show that B is both open and closed. If $y \in \mathrm{cls}\, B$, consider $x = p \circ f_1(y) = p \circ f_2(y)$ and U an elementary neighborhood containing x. The subset $(p \circ f_1)^{-1}(U) \cap (p \circ f_2)^{-1}(U)$ contains y. Because Y is locally connected, there is an open set W for which $y \in W \subset (p \circ f_1)^{-1}(U) \cap (p \circ f_2)^{-1}(U)$ with W connected. Then $f_1(W)$ and $f_2(W)$ are connected subsets of $p^{-1}(U) \subset \tilde{X}$. Since W is open and $y \in \mathrm{cls}\, B$, there is a point $z \in W$ with $z \in B$. Thus $f_1(z) = f_2(z)$ and $f_1(W) \cap f_2(W) \neq \emptyset$; therefore, $f_1(W)$ and $f_2(W)$ must lie in the same component of $p^{-1}(U)$. Since $p \circ f_1(y) = p \circ f_2(y)$ and the component in which we find both $f_1(y)$ and $f_2(y)$ is homeomorphic to U by the restriction of p, we have $f_1(y) = f_2(y)$. Thus $y \in B$ and B is closed.

If we let $y \in B$, the argument above shows that the sets $f_1(W)$ and $f_2(W)$ lie in the same component C_0 of $p^{-1}(U)$. It follows that, for all $w \in W$,

$$f_1(w) = (p|_{C_0})^{-1} \circ p \circ f_1(w) = (p|_{C_0})^{-1} \circ p \circ f_2(w) = f_2(w)$$

and so W is contained in B. Thus B is open.

The only subsets of Y that are both open and closed are Y itself and $\emptyset$ and so we have proved the lemma.   □

Using Lemma 8.3, two lifts of a path $\lambda \colon [0,1] \to X$ which begin at the same point in $\tilde{X}$ must be the same lift. This is the uniqueness part of Lemma 8.2.   □

Having lifted paths in X to paths in $\tilde{X}$, we next lift certain homotopies between paths.

Lemma 8.4. Let $p \colon \tilde{X} \to X$ be a covering space and let η_0, $\eta_1 \colon [0,1] \to \tilde{X}$ be two paths in $\tilde{X}$ with $\eta_0(0) = \eta_1(0) = \tilde{x}_0$. If $p \circ \eta_0(1) = x_1 = p \circ \eta_1(1)$ and $p \circ \eta_0 \simeq p \circ \eta_1$ via a homotopy that fixes the endpoints of the paths in X, then $\eta_1 \simeq \eta_2$ in $\tilde{X}$ and, in particular, $\eta_0(1) = \eta_1(1)$.

Proof. Let $H \colon [0,1] \times [0,1] \to X$ be a homotopy between $p \circ \eta_0$ and $p \circ \eta_1$. In this case, we have, for all $s, t \in [0,1]$,

$$H(s,0) = p \circ \eta_0(s) \quad\text{and}\quad H(0,t) = p(\tilde{x}_0)$$
$$H(s,1) = p \circ \eta_1(s) \qquad\qquad H(1,t) = p \circ \eta_0(1) = p \circ \eta_1(1).$$

Since $[0,1] \times [0,1]$ is a compact metric space, when we cover it by the collection $\{H^{-1}(U) \mid U,$ an elementary neighborhood of $X\}$, we can apply Lebesgue's Lemma to get $\delta > 0$ for which any subset of $[0,1] \times [0,1]$ of diameter $< \delta$ lies in some $H^{-1}(U)$. If we subdivide the interval $[0,1]$,

$$0 = s_0 < s_1 < \cdots < s_{m-1} < s_m = 1$$

and

$$0 = t_0 < t_1 < \cdots < t_{n-1} < t_n = 1,$$

so that $s_i - s_{i-1} < \delta/2$ and $t_j - t_{j-1} < \delta/2$, then H maps each subrectangle $[s_{i-1}, s_i] \times [t_{j-1}, t_j]$ into an elementary neighborhood for all i and j.

To construct the lifting $\hat{H}\colon [0,1] \times [0,1] \to \tilde{X}$ and show it is a homotopy between η_0 and η_1, begin by lifting H on $[0,s_1] \times [0,t_1]$ to $\tilde{X}$ by using $\hat{H} = (p|_{C_{11}})^{-1} \circ H$, where C_{11} is the component of $p^{-1}(U_{11})$ containing $\eta_0(0)$ and $H([0,s_1] \times [0,t_1]) \subset U_{11}$, an elementary neighborhood. Having done this, extend $\hat{H}$ next to $[s_1,s_2] \times [0,t_1]$. Notice that $\hat{H}$ has been defined on the line segment $\{s_1\} \times [0,t_1]$ which is connected and this determines the component of $p^{-1}(U_{21})$ for the elementary neighborhood U_{21} which contains $H([s_1,s_2] \times [0,t_1])$. Once the component, say C_{21}, is determined, extend $\hat{H}$ by $\hat{H} = (p|_{C_{21}})^{-1} \circ H$. Continue in this manner until $\hat{H}$ is defined on $[0,1] \times [0,t_1]$. Next, extend to $[0,1] \times [t_1,t_2]$ using the fact that the value of $\hat{H}$ has been determined on each successive subrectangle along the left and bottom edges, as a connected subset. Continue along each row until $\hat{H}$ is defined on $[0,1] \times [0,1]$. By Lemma 8.3, $\hat{H}$ is unique fulfilling the condition $\hat{H}(0,0) = \eta(0)$. Since $\eta_0(s)$ is also a lift of $H(s,0)$, we have that $\hat{H}(s,0) = \eta_0(s)$. The condition $H(0,t) = p \circ \eta_0(0)$ implies that $\hat{H}(0,t) = \eta_0(0)$, that is, the homotopy $\hat{H}$ is constant on the subset $\{0\} \times [0,1]$. Thus, the lift $\hat{H}(s,1)$ of the path $p \circ \eta_1(s)$ in X begins at $\eta_0(0) = \eta_1(0)$, and $\eta_1(s)$ is also such a lift. By uniqueness, $\hat{H}(s,1) = \eta_1(s)$. Finally, $H(1,t) = p \circ \eta_0(1) = p \circ \eta_1(1)$ for all $t \in [0,1]$, $\hat{H}(1,t) = \eta_0(1)$, and we conclude that $\eta_0(1) = \eta_1(1)$ since $\hat{H}(1,t)$ is constant. $\qquad\square$

Uniqueness of liftings of homotopies provides considerable control over the fundamental group through a covering space, giving us a toehold for computation.

Corollary 8.5. *Suppose* $p\colon \hat{X} \to X$ *is a covering space:* (1) *If* $\eta\colon [0,1] \to \tilde{X}$ *is a loop at* $\tilde{x}_0$ *and* $p \circ \eta$ *is homotopic to the constant loop* c_{x_0} *for* $x_0 = p(\tilde{x}_0)$, *then* $\eta \simeq c_{\tilde{x}_0}$. (2) *The induced homomorphism* $p_*\colon \pi_1(\tilde{X}, \tilde{x}_0) \to \pi_1(X, x_0)$ *is injective.* (3) *For all* $x \in X$, *the subsets* $p^{-1}(\{x\})$ *of* $\tilde{X}$ *have the same cardinality.*

Proof. (1) One lift of c_{x_0} is simply the constant path $c_{\tilde{x}_0}$. By Lemma 8.4, $p \circ \eta \simeq p \circ c_{\tilde{x}_0} = c_{x_0}$ implies $\eta \simeq c_{\tilde{x}_0}$.

(2) If $p_*([\lambda]) = p_*([\mu])$, then, because p_* is a homomorphism, $p_*([\lambda] * [\mu^{-1}]) = [c_{x_0}]$, that is, $p \circ (\lambda * \mu^{-1}) \simeq c_{x_0}$. By (1), $\lambda * \mu^{-1} \simeq c_{\tilde{x}_0}$ or $\lambda \simeq \mu$, that is, $[\lambda] = [\mu]$.

(3) Suppose x_0 and x_1 are in X and $\lambda\colon [0,1] \to X$ is a path joining x_0 to x_1. Suppose $y \in p^{-1}(\{x_0\})$. We define a mapping $\Lambda\colon p^{-1}(\{x_0\}) \to p^{-1}(\{x_1\})$ by lifting λ to $\lambda_y\colon [0,1] \to \tilde{X}$ with $\lambda_y(0) = y$. Define $\Lambda(y) = \lambda_y(1)$. Since λ_y is uniquely determined by being a lift of $p \circ \lambda_y = \lambda$ with $\lambda_y(0) = y$, the function Λ is well defined. By Lemma 8.3, lifts of λ beginning at different elements in $p^{-1}(\{x_0\})$ must end at different points in $p^{-1}(\{x_1\})$ and so Λ is injective. Using lifts of λ^{-1} we deduce that Λ is surjective. (Notice that a different choice of λ might give a different one-one correspondence Λ.) □

For $w\colon \mathbb{R} \to S^1$, $w(r) = e^{2\pi i r}$, we find that $w^{-1}(1 + 0i) = \mathbb{Z} \subset \mathbb{R}$ and so $w^{-1}(\{z\})$ is countably infinite for each $z \in S^1$. For $p\colon S^2 \to \mathbb{RP}^2$, $p^{-1}(\{[\pm\mathbf{x}_0]\})$ contains two elements, $\mathbf{x}_0$ and $-\mathbf{x}_0$. In general, if we lift a loop $\omega\colon [0,1] \to X$ at x_0 in X, the proof of (3) of Corollary 8.5 obtains a mapping $\Omega\colon p^{-1}(\{x_0\}) \to p^{-1}(\{x_0\})$ by lifting the loop. By remark (1) of the corollary, if Ω is nontrivial, then the loop ω is not homotopic to the constant map. This observation is enough to prove the following.

Theorem 8.6. A. $\pi_1(S^1) \cong \mathbb{Z}$. **B.** $\pi_1(\mathbb{RP}^2) \cong \mathbb{Z}/2\mathbb{Z}$.

Proof of A. If $\beta\colon [0,1] \to S^1$ is any loop at $1 \in S^1$, then the lift of β to $\hat{\beta}\colon [0,1] \to \mathbb{R}$ satisfies $\hat{\beta}(1) \in \mathbb{Z}$. The properties of liftings determine a function $\Xi\colon \pi_1(S^1) \to \mathbb{Z}$ given by $[\beta] \mapsto \hat{\beta}(1)$.

Let $\alpha\colon [0,1] \to S^1$ be given by $\alpha(t) = (\cos(2\pi t), \sin(2\pi t))$. Since $\alpha = w|_{[0,1]}$, we see that one lift of α to $\mathbb{R}$ is just the identity and $\hat{\alpha}(1) = 1$. It follows that α is not homotopic to the constant map at 1, c_1. Next consider α^n for $n \in \mathbb{Z}$, given by $\alpha^n(t) = (\cos(2\pi n t), \sin(2\pi n t))$. By the same argument for α, $\hat{\alpha}^n(1) = n$ and so the mapping $\Xi\colon \pi_1(S^1) \to \mathbb{Z}$ is surjective. Since each $\alpha^n \not\simeq c_1$ for $n \neq 0$, the subgroup generated by $[\alpha]$, isomorphic to $\mathbb{Z}$, is a subgroup of $\pi_1(S^1)$.

To finish the proof of **A**, we show that if β is any loop based at 1 in S^1, then $\beta \simeq \alpha^n$ for some $n \in \mathbb{Z}$. Let $U_1 = \{(x,y) \in S^1 \mid y > -1/10\}$ and $U_2 = \{(x,y) \in S^1 \mid y < 1/10\}$. The pair $\beta^{-1}(U_1)$, $\beta^{-1}(U_2)$ is an open cover of $[0,1]$ and by Lebesgue's Lemma we can subdivide $[0,1]$ as $0 = t_0 < t_1 < \cdots < t_{m-1} < t_m = 1$ so that

i) $\beta([t_i, t_{i+1}]) \subset U_1$ or $\beta([t_i, t_{i+1}]) \subset U_2$ for $0 \leq i < m$.

Form the union of consecutive subintervals when both are mapped to the same U_j, $j = 1$ or 2. In detail, let $s_0 = 0$ and $s_1 = t_{i_1}$, where $\beta([0, t_{i_1}]) \subset U_{j_1}$ for j_1 one of 1 or 2 and $\beta([t_{i_1}, t_{i_1+1}]) \not\subset U_{j_1}$. Let $U_{j_2} \neq U_{j_1}$ and $\beta([t_{i_1}, t_{i_1+1}]) \subset U_{j_2}$. Then let $s_2 = t_{i_2}$, where $\beta([t_{i_1}, t_{i_2}]) \subset U_{j_2}$ but $\beta([t_{i_2}, t_{i_2+1}]) \not\subset U_{j_2}$. Continue in this manner to get

$$0 = s_0 < s_1 < \cdots < s_{k-1} < s_k = 1$$

so that

ii) $\beta([s_{j-1}, s_j])$ and $\beta([s_j, s_{j+1}])$ are not both contained in the same U_k, for $k = 1, 2$.

Let $\beta_j \colon [0, 1] \to S^1$ denote the reparameterization of $\beta|_{[s_j, s_{j+1}]}$ so that $\beta \simeq \beta_0 * \beta_1 * \cdots * \beta_{k-1}$, and each β_j is a path in exactly one of U_1 or U_2. Furthermore, $\beta(s_j) \in U_1 \cap U_2$, a subspace of two components, one of which contains $1 = e^{2\pi i 0}$ and the other $-1 = e^{\pi i}$. For $0 < j < m$ choose a path $\lambda_j \colon [0, 1] \to U_1 \cap U_2$ with $\lambda_j(0) = \beta(s_j) = \beta_{j-1}(s_j)$ and $\lambda_j(1) = 1$ or -1, depending on the component. Define

$$\gamma_1 = \beta_0 * \lambda_1,$$
$$\gamma_j = \lambda_{j-1}^{-1} * \beta_{j-1} * \lambda_j \text{ for } 1 < j < k,$$
$$\gamma_k = \lambda_{m-1}^{-1} * \beta_{k-1}.$$

By cancelling $\lambda_j * \lambda_j^{-1}$, $\beta \simeq \gamma_1 * \gamma_2 * \cdots * \gamma_k$. Consider the paths γ_ℓ. If γ_ℓ is a closed path, it lies in U_1 or U_2, which are simply-connected and so $\gamma_\ell \simeq c_1$ or $\gamma_\ell \simeq c_{-1}$. If γ_ℓ is not closed, then it crosses between the components of $U_1 \cap U_2$. It follows that $\gamma_\ell \simeq \eta_1^{\pm 1}$ or $\gamma_\ell \simeq \eta_2^{\pm 1}$, where $\eta_1(t) = (\cos(\pi t), \sin(\pi t))$, the path joining 1 to -1 in U_1, and $\eta_2(t) = (\cos(\pi t + \pi), \sin(\pi t + \pi))$, the path joining -1 to 1 in U_2. Making the cancellations of the type $\eta_1 \eta_1^{-1} \simeq c_1$ or $\eta_2 \eta_2^{-1} \simeq c_{-1}$, we are left with three possibilities:

$$\beta \simeq c_1, \quad \beta \simeq \eta_1 * \eta_2 * \eta_1 * \eta_2 * \cdots * \eta_1 * \eta_2, \text{ or}$$
$$\beta \simeq \eta_2^{-1} * \eta_1^{-1} * \eta_2^{-1} * \cdots * \eta_2^{-1} * \eta_1^{-1},$$

after cancelling out $c_{\pm 1}$. The ordering is determined by the fact that β begins and ends at 1, and each γ_ℓ either joins 1 to -1, joins -1 to 1, or it simply stays put. After cancellation of the paths that stay put or products of paths that are homotopic to the constant path, we are left with such a product in that order. Finally, $w|_{[0,1]} = \alpha \simeq \eta_1 * \eta_2$ and so $\beta \simeq \alpha^n$ for some $n \in \mathbb{Z}$. $\qquad\square$

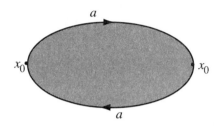

Proof of B. Consider the model of the projective plane given by the *di-gon*, a disk with each point on the boundary identified with the point symmetric with respect to the origin. Let $x_0 \in \mathbb{R}\mathrm{P}^2$ be the point $x_0 = [\pm(1,0,0)]$. Let $p\colon S^2 \to \mathbb{R}\mathrm{P}^2$ denote the covering space $p(\mathbf{x}) = [\pm\mathbf{x}]$. Let the loop a in $\mathbb{R}\mathrm{P}^2$ denote *half of the equator*, and lift a to S^2. We get a path $\hat{a}$ from $(1,0,0)$ to $(-1,0,0)$ along the equator of S^2. By Corollary 8.5, $a \not\simeq c_{x_0}$. In the di-gon representation of $\mathbb{R}\mathrm{P}^2$, $a * a = a^2$ surrounds the disk, and so a^2 can be contracted to c_{x_0} by shrinking to the center of the disk and translating over to x_0. It follows that $\pi_1(\mathbb{R}\mathrm{P}^2)$ contains $\mathbb{Z}/2\mathbb{Z}$. To finish, we need show that any loop at x_0 is homotopic to a^n for some $n \in \mathbb{Z}$. Using the di-gon we see that away from the image of the path a^2 a path lies in the contractible interior of a disk. The disk can be used to push any loop onto a as often as it crosses between the copies of x_0. Thus we see that any loop based at x_0 is homotopic to a^n for some $n \in \mathbb{Z}$ and so homotopic to a or c_{x_0}. This implies that

$$\pi_1(\mathbb{R}\mathrm{P}^2) = \langle [a] \rangle / ([a]^2 = [c_{x_0}]) \cong \mathbb{Z}/2\mathbb{Z}.$$

This completes the proof of Theorem 8.6. $\qquad\square$

Covering spaces can be developed much further. We refer the reader to [**55**] or [**51**] for thorough treatments. Let's turn now to applications. We first return to the central question of the text:

Invariance of Dimension for $(2, n)$**.** *For* $n \neq 2$, $\mathbb{R}^n$ *and* $\mathbb{R}^2$ *are not homeomorphic.*

Proof. We assume that $n \geq 2$ since the case of $n = 1$ is covered in Chapter 5. If $\mathbb{R}^n \cong \mathbb{R}^2$, then, by composing with a translation if needed, we can choose a homeomorphism $f \colon \mathbb{R}^n \to \mathbb{R}^2$ for which $f(\mathbf{0}) = (0,0)$. Such a mapping induces a homeomorphism $\mathbb{R}^n - \{\mathbf{0}\} \cong \mathbb{R}^2 - \{(0,0)\}$. Since S^{n-1} is a deformation retract of $\mathbb{R}^n - \{\mathbf{0}\}$, by Theorem 7.10, $\pi_1(\mathbb{R}^n - \{\mathbf{0}\}) \cong \pi_1(S^{n-1})$. For $n > 2$, Corollary 7.13 states that $\pi_1(S^{n-1}) \cong \{e\}$, while, for $n = 2$, $\pi_1(S^1) \cong \mathbb{Z}$. Since the fundamental group is a topological invariant, it must be the case that $n = 2$. $\qquad\square$

This argument is characteristic of the power of introducing algebraic structures as topological invariants of spaces. Our goal in later chapters is to generalize these ideas.

Recall the somewhat unexpected topological property introduced in the exercises of Chapter 2: A space X has the *fixed point property* (**FPP**) if any continuous mapping $f \colon X \to X$ has a fixed point, that is, there exists a point $x_0 \in X$ with $f(x_0) = x_0$. By the Intermediate Value Theorem we can prove that the interval $[0,1]$ has the FPP: if $f \colon [0,1] \to [0,1]$ is continuous, then define $g(x) = f(x) - x \colon [0,1] \to \mathbb{R}$. If $f(0) \neq 0$ and $f(1) \neq 1$, then $g(0) > 0$ and $g(1) < 0$ and so g must take the value 0 somewhere. If $g(x) = 0$, then $f(x) = x$.

What is the generalization of the space $[0, 1]$ to higher dimensions? Candidates include $[0,1] \times [0,1]$ in dimension 2 or maybe the **two-disk** $e^2 = \{\mathbf{x} \in \mathbb{R}^2 \mid \|\mathbf{x}\| \leq 1\} = \operatorname{cls} B(\mathbf{0}, 1)$. The choice between these two candidates is irrelevant since the fixed point property is a topological property and they are homeomorphic. (Can you prove it?) We generalize the fixed point property for the interval $[0, 1]$ to the two-disk.

Theorem 8.7 (Brouwer's Theorem in dimension 2). *The two-disk $e^2 = \{\mathbf{x} \in \mathbb{R}^2 \mid \|\mathbf{x}\| \leq 1\} \subset \mathbb{R}^2$ has the fixed point property.*

Proof. Suppose $f \colon e^2 \to e^2$ is a continuous function without a fixed point. Then for each $\mathbf{x} \in e^2$, $f(\mathbf{x}) \neq \mathbf{x}$. Define $g \colon e^2 \to S^1$ by

$$g(\mathbf{x}) = \text{intersection of the ray from } f(\mathbf{x}) \text{ to } \mathbf{x} \text{ with } S^1.$$

To see that $g(\mathbf{x})$ is continuous on e^2, we apply some vector geometry: write $Q = f(\mathbf{x})$, $Z = g(\mathbf{x})$. Let $O = (0,0)$ and define

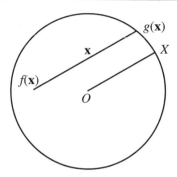

$X = (\mathbf{x} - Q)/\|\mathbf{x} - Q\|$. Then, $g(\mathbf{x}) = Z = Q + tX$ for some $t \geq 0$ for which $Q + tX \in S^1$, that is, $(Q + tX) \cdot (Q + tX) = 1$. This condition can be rewritten to solve for t, namely,

$$(Q + tX) \cdot (Q + tX) = t^2(X \cdot X) + 2t(Q \cdot X) + Q \cdot Q = 1.$$

The quadratic formula gives

$$t_{\mathbf{x}} = -Q \cdot X + \sqrt{(Q \cdot X)^2 + 1 - Q \cdot Q}$$

$$= -f(\mathbf{x}) \cdot \frac{\mathbf{x} - f(\mathbf{x})}{\|\mathbf{x} - f(\mathbf{x})\|} + \sqrt{\left(f(\mathbf{x}) \cdot \frac{\mathbf{x} - f(\mathbf{x})}{\|\mathbf{x} - f(\mathbf{x})\|}\right)^2 + 1 - f(\mathbf{x}) \cdot f(\mathbf{x})}.$$

Note that this choice of signs gives $t_{\mathbf{x}} \geq 0$, and $t_{\mathbf{x}} = 0$ implies $f(\mathbf{x}) = \mathbf{x}$. Since we have assumed that this doesn't happen, $t_{\mathbf{x}} > 0$. Furthermore, $t_{\mathbf{x}}$ is a continuous function of $\mathbf{x}$. We can write $g(\mathbf{x})$ explicitly as

$$g(\mathbf{x}) = f(\mathbf{x}) + t_{\mathbf{x}} \frac{\mathbf{x} - f(\mathbf{x})}{\|\mathbf{x} - f(\mathbf{x})\|}$$

and so $g(\mathbf{x})$ is continuous.

By the definition of the mapping g, if $\mathbf{x} \in S^1 \subset e^2$, then $g(\mathbf{x}) = \mathbf{x}$. We have constructed a continuous mapping $g \colon e^2 \to S^1$ for which $g \circ i = \mathrm{id}_{S^1}$, that is, the identity mapping on S^1 can be factored:

$$\mathrm{id}_{S^1} \colon S^1 \xrightarrow{\ i\ } e^2 \xrightarrow{\ g\ } S^1.$$

This composite leads to a composite of group homomorphisms and fundamental groups:

$$\mathrm{id} \colon \pi_1(S^1) \xrightarrow{\ i_*\ } \pi_1(e^2) \xrightarrow{\ g_*\ } \pi_1(S^1).$$

However, $\pi_1(e^2) = \{[c_1]\}$ and so $g_* \circ i_*([\alpha]) = [c_1] \neq [\alpha]$ and $g_* \circ i_* \neq$ id, a contradiction. Therefore, a continuous function $f \colon e^2 \to e^2$ without fixed points is not possible. $\qquad\square$

Corollary 8.8. S^1 *is not a retract of* e^2.

More powerful tools will be developed in later chapters to prove a generalization of Theorem 8.7 and its corollary. Brouwer proved this general result around 1911 [11].

We next apply the fact that $\pi_1(\mathbb{RP}^2) \cong \mathbb{Z}/2\mathbb{Z}$. Recall that $\mathbb{RP}^2$ is the space of lines through the origin in $\mathbb{R}^3$. The lower dimensional analogue is the space $\mathbb{RP}^1$ consisting of lines through the origin in $\mathbb{R}^2$. We can identify a line with the angle it makes with the x-axis. To obtain every line through the origin, we only need angles $0 \leq \theta \leq \pi$, where the x-axis is identified with the angles 0 and π. Hence $\mathbb{RP}^1 \cong [0, \pi]/(0 \sim \pi) \cong S^1$. Thus $\pi_1(\mathbb{RP}^1) \cong \mathbb{Z}$. The analogue of the covering map $p \colon S^2 \to \mathbb{RP}^2$ in this case is $\overline{p} \colon S^1 \to \mathbb{RP}^1$ given by $e^{2\pi i\theta} \mapsto [\pm e^{2\pi i\theta}]$. In fact, $\overline{p}_* \colon \pi_1(S^1) \to \pi_1(\mathbb{RP}^1)$ is described as a homomorphism $\mathbb{Z} \to \mathbb{Z}$ given by multiplication by two, because the generator $[\alpha]$ wraps around $\mathbb{RP}^1$ twice.

In Chapter 5 we proved that a continuous mapping $f \colon S^1 \to \mathbb{R}$ must send some point and its negative to the same value, that is, there is always a point $x_0 \in S^1$ with $f(x_0) = f(-x_0)$. We can generalize that result to S^2.

Theorem 8.9. *If* $f \colon S^2 \to \mathbb{R}^2$ *is a continuous function, then there exists a point* $\mathbf{x} \in S^2$ *with* $f(\mathbf{x}) = f(-\mathbf{x})$.

We proceed by proving an associated result.

Proposition 8.10 (The Borsuk-Ulam Theorem for $n = 2$). *There does not exist a continuous function* $f \colon S^2 \to S^1$ *that satisfies* $f(-\mathbf{x}) = -f(\mathbf{x})$ *for all* $\mathbf{x} \in S^2$.

Proof of the Borsuk-Ulam Theorem. Assume such a function exists. The condition satisfied by f can be written $f(\pm\mathbf{x}) = \pm f(\mathbf{x})$.

It follows that f induces $\hat{f}\colon \mathbb{R}P^2 \to \mathbb{R}P^1$ and $\hat{f}$ fits into a diagram:

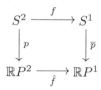

for which $\bar{p} \circ f = \hat{f} \circ p$. Consider the induced homomorphism $\hat{f}_*$: $\pi_1(\mathbb{R}P^2) \to \pi_1(\mathbb{R}P^1)$. By Theorem 8.6, $\hat{f}_*$ is a homomorphism $\mathbb{Z}/2\mathbb{Z} \to \mathbb{Z}$. However, any such homomorphism must be the trivial homomorphism. (Do you know why?) Let $\lambda\colon [0,1] \to S^2$ denote a path from the North Pole to the South Pole along a meridian of constant longitude. It follows that $[p \circ \lambda] = [\alpha]$, a generator for $\mathbb{Z}/2\mathbb{Z} \cong \pi_1(\mathbb{R}P^2)$. Since the North and South Poles are antipodal, these points are identified in $\mathbb{R}P^1$ after passage through f and $\bar{p}$. Hence $[\bar{p} \circ f \circ \lambda]$ is nontrivial in $\pi_1(\mathbb{R}P^1)$. But $[\bar{p} \circ f \circ \lambda] = [\hat{f} \circ p \circ \lambda] = \hat{f}_*([p \circ \lambda]) = 0$, a contradiction. $\square$

Corollary 8.11. *If $f\colon S^2 \to \mathbb{R}^2$ is a continuous function such that $f(-\mathbf{x}) = -f(\mathbf{x})$ for all $\mathbf{x} \in S^2$, then $f(\mathbf{x}) = (0,0)$ for some $\mathbf{x} \in S^2$.*

Proof. If not, then $g(\mathbf{x}) = f(\mathbf{x})/\|f(\mathbf{x})\|$ would be a continuous function $g\colon S^2 \to S^1$ with $g(-\mathbf{x}) = -g(\mathbf{x})$ for all $\mathbf{x} \in S^2$. $\square$

Proof of Theorem 8.9. Suppose for every $\mathbf{x} \in S^2$ that $f(\mathbf{x}) \neq f(-\mathbf{x})$. Then define $g(\mathbf{x}) = f(\mathbf{x}) - f(-\mathbf{x})$. Notice that g is continuous, $g(-\mathbf{x}) = -g(\mathbf{x})$, and $g(\mathbf{x}) \neq \mathbf{0}$ for all $\mathbf{x} \in S^2$, a contradiction. $\square$

Corollary 8.12. *No subset of $\mathbb{R}^2$ is homeomorphic to S^2.*

The corollary tells us that there is no cartographic map homeomorphic to the entire sphere.

Finally, we derive an unexpected corollary of our analysis of the fundamental group of the circle, namely, the Fundamental Theorem of Algebra. This topological proof gives a complete proof avoiding the difficulties in the approach of Gauss in Chapter 5 based on connectedness.

The Fundamental Theorem of Algebra. *If* $p(z) = z^n + a_{n-1}z^{n-1}$ $+ \cdots + a_1 z + a_0$ *is a polynomial with complex coefficients, then there is a complex number z_0 with $p(z_0) = 0$.*

Proof. Recall that $\mathbb{C} \cong \mathbb{R}^2$ and the nth power mapping $h \colon z \mapsto z^n$ induces a mapping $h \colon S^1 \to S^1$ which can be written as $e^{i\theta} \mapsto e^{in\theta}$. Lifting this mapping to the covering space $w \colon \mathbb{R} \to S^1$, it represents $n \cdot \in \mathbb{Z} \cong \pi_1(S^1)$ via the identification of $\pi_1(S^1)$ with $\mathbb{Z}$ given by $[\beta] \mapsto \hat{\beta}(1)$.

Viewed as a mapping, $h \colon S^1 \to S^1$, h induces the homomorphism $h_* \colon \pi_1(S^1) \to \pi_1(S^1)$. The law of exponents implies that

$$h_*(\theta \mapsto e^{\pi i m\theta}) = (\theta \mapsto (e^{\pi i m\theta})^n = e^{\pi i n m\theta}),$$

that is, h_* is multiplication by n.

We first consider a special case of the theorem—suppose

$$|a_{n-1}| + |a_{n-2}| + \cdots + |a_0| < 1.$$

Suppose $p(z)$ has no root in $e^2 = \{z \in \mathbb{C} \mid |z| \leq 1\}$. Define the mapping $\hat{p} \colon e^2 \to \mathbb{R}^2 - \{\mathbf{0}\}$ by $\hat{p}(z) = p(z)$. Restricting to $S^1 = \partial e^2$ we get $\hat{p}| \colon S^1 \to \mathbb{R}^2 - \{\mathbf{0}\}$. Since $\hat{p}|$ can be extended to e^2, it follows (exercise) that $\hat{p}|$ is homotopic to a constant map. However, consider the mapping

$$F(z,t) = z^n + t(a_{n-1}z^{n-1} + \cdots + a_0),$$

which gives a homotopy between $F(z,0) = z^n$ and $F(z,1) = p(z)$. If $F(z,t)$ never vanishes on S^1, the homotopy implies $\hat{p}| \simeq z^n$. To establish this condition, for $|z| = 1$ we estimate

$$\begin{aligned} |F(z,t)| &\geq |z^n| - |t(a_{n-1}z^{n-1} + \cdots + a_0)| \\ &\geq 1 - t(|a_{n-1}z^{n-1}| + \cdots + |a_0|) \\ &= 1 - t(|a_{n-1}| + \cdots + |a_0|) > 0. \end{aligned}$$

As a class in $\pi_1(S^1)$, $[(z \mapsto z^n)]$ is not homotopic to the constant map while $\hat{p}|$ is, so we get a contradiction.

To reduce the general case to this special case, let $t \in \mathbb{R}$, $t \neq 0$, and let $u = tz$. So

$$p(u) = u^n + a_{n-1}u^{n-1} + \cdots + a_1 u + a_0$$
$$= (tz)^n + a_{n-1}(tz)^{n-1} + \cdots + a_1 tz + a_0.$$

If $p(u) = 0$, then

$$z^n + \frac{a_{n-1}}{t}z^{n-1} + \cdots + \frac{a_1}{t^{n-1}}z + \frac{a_0}{t^n} = 0.$$

So given a zero for $p(u)$ we get a zero for $\tilde{p}_t(z)$ with $\tilde{p}_t(z) = z^n + \frac{a_{n-1}}{t}z^{n-1} + \cdots + \frac{a_0}{t^n}$ and vice versa. Taking t large enough we can guarantee

$$\left|\frac{a_{n-1}}{t}\right| + \cdots + \left|\frac{a_1}{t^{n-1}}\right| + \left|\frac{a_0}{t^n}\right| < 1$$

and we can apply the special case. $\square$

In Chapter 7 we proved that a subspace A of a space X, which is a deformation retract of X, shares the same fundamental group as X. Furthermore, if X and Y are homeomorphic spaces, they share the same fundamental group. We generalize these conditions to identify an important relation between spaces.

Definition 8.13. Two spaces are **homotopy equivalent**, denoted $X \simeq Y$, if there are mappings $f: X \to Y$ and $g: Y \to X$ with $g \circ f \simeq \mathrm{id}_X$ and $f \circ g \simeq \mathrm{id}_Y$.

If $A \subset X$ is a deformation retract, then there is a mapping $r: X \to A$ for which $\mathrm{id}_A = r \circ i: A \to A$ and $\mathrm{id}_X \simeq i \circ r: X \to X$. Thus A is homotopy equivalent to X and homotopy equivalence generalizes the relation of deformation retraction. Contractible spaces are homotopy equivalent to a one-point space so homotopy equivalence is a weaker notion than homeomorphism.

Proposition 8.14. *In a set of topological spaces, homotopy equivalence is an equivalence relation.*

Proof. It suffices to check transitivity since the other properties are clear. Suppose $X \simeq Y$ and $Y \simeq Z$ via mappings $f: X \to Y$, $g: Y \to$

$X, t\colon Y \to Z$, and $u\colon Z \to Y$. Consider $t \circ f\colon X \to Z$ and $g \circ u\colon Z \to X$. Then

$$(g \circ u) \circ (t \circ f) \simeq g \circ (u \circ t) \circ f \simeq g \circ \mathrm{id}_Y \circ f = g \circ f \simeq \mathrm{id}_X \text{ and}$$
$$(t \circ f) \circ (g \circ u) \simeq t \circ (f \circ g) \circ u \simeq t \circ \mathrm{id}_X \circ u = t \circ u \simeq \mathrm{id}_Z \,.$$

Fixing a universe, that is, a set in which all relevant spaces are elements, the equivalence class of a space X is called its **homotopy type**. The effectiveness of the fundamental group to distinguish spaces is limited by homotopy equivalence.

Proposition 8.15. *If X and Y are homotopy-equivalent spaces via mappings $f\colon X \to Y$ and $g\colon Y \to X$, then the induced mappings $f_*\colon \pi_1(X, x_0) \to \pi_1(Y, f(x_0))$ and $g_*\colon \pi_1(Y, y_0) \to \pi_1(X, g(y_0))$ are isomorphisms.*

Proof. Let $H\colon X \times [0, 1] \to X$ be a homotopy between $g \circ f$ and id_X. Let $\gamma\colon [0, 1] \to X$ be the path $\gamma(t) = H(x_0, t)$, so that $\gamma(0) = g \circ f(x_0)$ and $\gamma(1) = x_0$. We can write the induced homomorphisms:

$$\pi_1(X, x_0) \xrightarrow{f_*} \pi_1(Y, f(x_0)) \xrightarrow{g_*} \pi_1(X, g \circ f(x_0)) \xrightarrow{u_\gamma} \pi_1(X, x_0).$$

We claim that this composite is the identity homomorphism. Consider $[\lambda] \in \pi_1(X, x_0)$. The result of the composite on this element is

$$[\lambda] \mapsto [f \circ \lambda] \mapsto [g \circ f \circ \lambda] \mapsto [\gamma^{-1} * (g \circ f \circ \lambda) * \gamma].$$

Apply the homotopy H to get a homotopy from $g \circ f \circ \lambda$ to λ by $H(\lambda(t), s)$. We use this homotopy to construct one from $\gamma^{-1} * (g \circ f \circ \lambda) * \gamma$ to λ by reparameterizing according to the diagram:

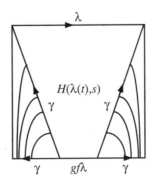

In the triangles, we have taken γ and opened it into a triangle with the pictured curves given by isobars (constant paths). It follows from the homotopy that $[\gamma^{-1} * (g \circ f \circ \lambda) * \gamma] = [\lambda]$. This implies that f_* is injective and g_* surjective. To finish the proof consider the composite

$$\pi_1(Y, f(x_0)) \xrightarrow{g_*} \pi_1(X, g \circ f(x_0))$$

$$\xrightarrow{f_*} \pi_1(Y, f \circ g \circ f(x_0)) \xrightarrow{u_\eta} \pi_1(Y, f(x_0)),$$

where $\eta\colon [0,1] \to Y$ is the path $\eta(t) = \bar{H}(f(x_0), t)$ in the homotopy $\bar{H}$ between $f \circ g$ and id_Y. The same argument applies *mutatis mutandis* to show that f_* is surjective and g_* is injective and hence both homomorphisms are isomorphisms.   □

Homotopy equivalence is cruder than homeomorphism but includes it as a special case. To give an idea of how crude homotopy equivalence is, notice that, for all n, $\mathbb{R}^n$ is homotopy equivalent to a point. The letters of the alphabet as subspaces of $\mathbb{R}^2$ show other failures to distinguish between different topological spaces.

$$A \simeq D \simeq S^1, B \simeq S^1 \vee S^1, C \simeq E \simeq F \simeq *, \ldots.$$

Proposition 8.15 shows that the fundamental group is a **homotopy invariant**, that is, if $X \simeq Y$, then $\pi_1(X) \cong \pi_1(Y)$. Thinking of the fundamental group as a filter that distinguishes spaces, it can only hope to catch homotopy inequivalent spaces. In later chapters we will consider other homotopy invariants. Poincaré [66] introduced the fundamental group to distinguish certain manifolds that were indistinguishable via other combinatorial invariants.

Exercises

1. Suppose that $f\colon S^1 \to S^1$ has an extension $\hat{f}\colon e^2 \to S^1$, that is, the mapping $\hat{f}$ satisfies $\hat{f} \circ i = f$, where $i\colon S^1 \to e^2$ is the inclusion. Show that f is **null-homotopic**, that is, f is homotopic to the constant mapping.

2. Though we will not prove it, one of the useful theorems for computing the fundamental groups of spaces is the **Seifert-vanKampen Theorem**. A special case of this theorem

is the following: *If a path-connected space X is a union $X = U \cup V$ with V simply-connected and $x_0 \in U \cap V$, then the inclusion $i \colon U \to X$ induces a surjection $i_* \colon \pi_1(U, x_0) \to \pi_1(X, x_0)$ with kernel given by the smallest subgroup of $\pi_1(U, x_0)$ containing $j_*(\pi_1(U \cap V, x_0))$, where $j \colon U \cap V \hookrightarrow U$ denotes the inclusion.* Use the descriptions of $\mathbb{RP}^2$ of previous chapters and this theorem to make another computation of $\pi_1(\mathbb{RP}^2)$.

3. Suppose that X is simply-connected and $p \colon \tilde{X} \to X$ is a covering space of X. Show that p is a homeomorphism.

4. Let $\Omega(X, x_0)$ denote the based loop space of X given by

$$\Omega(X, x_0) = \{\lambda \colon [0, 1] \to X \mid \lambda \text{ is continuous and } \lambda(0) = \lambda(1) = x_0\}.$$

This subspace of $\mathrm{map}(I, X)$ is topologized with the compact-open topology. Show that

   i) $\pi_0(\Omega(X, x_0))$, the collection of path-components of $\Omega(X, x_0)$, is in one-one correspondence with $\pi_1(X, x_0)$.

   ii) Show that the loop multiplication $m \colon \Omega(X, x_0) \times \Omega(X, x_0) \to \Omega(X, x_0)$ given by $m(\lambda, \mu) = \lambda * \mu$ is a continuous multiplication on $\Omega(X, x_0)$.

5. We know from Theorem 7.15 and Theorem 8.6 that the fundamental group of the torus $S^1 \times S^1$ is $\mathbb{Z} \times \mathbb{Z}$. Use the argument for the computation of $\pi_1(\mathbb{RP}^2) \cong \mathbb{Z}/2\mathbb{Z}$ to prove $\pi_1(S^1 \times S^1) \cong \mathbb{Z} \times \mathbb{Z}$ by viewing the torus as a quotient of $[0, 1] \times [0, 1]$.

6. Let's make a space—take two distinct 2-spheres, S^2, and join them by a line segment—a space resembling dumbbells, but with a very thin connector. Denote this space by X and show that it is simply-connected.

Chapter 9

The Jordan Curve Theorem

> *It is established then that every continuous (closed) curve divides the plane into two regions, one exterior, one interior, ...*

CAMILLE JORDAN, 1882

In his 1882 *Cours d'analyse* [**41**], CAMILLE JORDAN (1838–1922) stated a classical theorem, topological in nature and inadequately proved by Jordan. The theorem concerns separation and connectedness in the plane on one hand, and the topological properties of simple, closed curves on the other.

The Jordan Curve Theorem. *If C is a simple, closed curve in the plane $\mathbb{R}^2$, that is, $C \subset \mathbb{R}^2$ and C is homeomorphic to S^1, then $\mathbb{R}^2 - C$, the complement of C, has two components, each sharing C as boundary.*

The statement of the theorem borders on the obvious—few would doubt it to be true. However, mathematicians of the nineteenth century had developed a healthy respect for the monstrous possibilities that their new researches into analysis revealed. Furthermore, a proof using rigorous and appropriate tools of a fact that seemed obvious meant that the obvious was a solid point of departure for generalization.

The proof that follows is an amalgam of two celebrated proofs— the principal part is based on work of Brouwer in which the notion of the *index* of a point relative to a curve plays a key role. Brouwer's proof was simplified by ERHARD SCHMIDT (1876–1959) (see [72] and [6]). The second proof, due to J. W. ALEXANDER (1888–1971) is based on the combinatorial and algebraic notion of a *grating* (see [63]). Although each proof can be developed independently, the main ideas of combinatorial approximation and an index provide a framework for generalizations that will be the focus of the final chapters.

A **Jordan curve**, or *simple, closed curve*, is a subset C of $\mathbb{R}^2$ that is homeomorphic to a circle. A **Jordan arc**, or *simple arc*, is a subset of $\mathbb{R}^2$ homeomorphic to a closed line segment in $\mathbb{R}$. A choice of homeomorphism gives a parameterization of the Jordan curve or arc, $\alpha\colon [0,1] \to \mathbb{R}^2$, as the composite of the homeomorphism $f\colon S^1 \to C \subset \mathbb{R}^2$ with $w\colon [0,1] \to S^1$, given by $w(t) = (\cos 2\pi t, \sin 2\pi t)$. A Jordan curve will have many choices of parameterization α and so relevant properties of the curve C must be shown to be independent of the choice of α. Notice that the parameterization $\alpha\colon [0,1] \to \mathbb{R}^2$ is one-one on $[0,1)$ and $\alpha(0) = \alpha(1)$.

Gratings and arcs

We begin by analyzing the separation properties of Jordan arcs. Choose a homeomorphism $\lambda\colon [0,1] \to \Lambda \subset \mathbb{R}^2$, which parameterizes the arc. Notice that $\Lambda = \lambda([0,1])$ is compact and closed in $\mathbb{R}^2$ and so $\mathbb{R}^2 - \Lambda$ is open.

Separation Theorem for Jordan arcs. *A Jordan arc Λ does not separate the plane, that is, $\mathbb{R}^2 - \Lambda$ is connected.*

Since $\mathbb{R}^2$ is locally path-connected, the complement of Λ is connected if and only if it is path-connected. An intuitive argument to establish the separation theorem begins with a pair of points P and Q in $\mathbb{R}^2 - \Lambda$. We can join P and Q by a path in $\mathbb{R}^2$, and then try to show that the path can be modified to a path that avoids Λ. It may be the case that Λ is very complicated, and a general proof requires great care to show that you can always find such a path.

Toward a rigorous argument we introduce a combinatorial structure that will allow us to make the modifications of paths in a methodical manner and so turn intuition into proof. The combinatorial structure is interesting in its own right—it combines approximation and algebraic manipulation, features that will be generalized to spaces in the remaining chapters. It is the interplay between the topological and combinatorial that makes this structure so useful. I have followed the classic text of Newman [**63**] in this section.

A *square region* in the plane is a subset $S = [a, a+s] \times [b, b+s] \subset \mathbb{R}^2$, where $a, b \in \mathbb{R}$ and $s > 0$. The region may be subdivided into rectangles by choosing values

$$a = a_0 < a_1 < a_2 < \cdots < a_{n-1} < a_n = a + s,$$
$$b = b_0 < b_1 < b_2 < \cdots < b_{m-1} < b_m = b + s,$$

with subrectangles given by $[a_i, a_{i+1}] \times [b_j, b_{j+1}]$ for $0 \le i < n$ and $0 \le j < m$. Such a subdivision is called a **grating**, introduced by Alexander in [**3**]. We denote a grating by $\mathcal{G} = (S, \{a_i\}, \{b_j\})$.

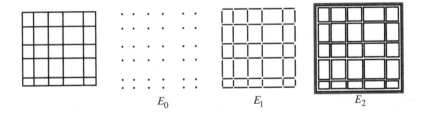

$$E_0 \qquad\qquad E_1 \qquad\qquad E_2$$

To a grating $\mathcal{G}$ we associate the following combinatorial data:

i) $E_0(\mathcal{G}) = \{(a_i, b_j) \in \mathbb{R}^2 \mid 0 \le i \le n, 0 \le j \le m\}$, its set of *vertices or 0-cells*;

ii) $E_1(\mathcal{G}) = \{PQ \mid P = (a_i, b_j) \text{ and } Q = (a_{i+1}, b_j) \text{ or } (a_i, b_{j+1})\}$, its set of *edges or 1-cells*, and

iii) $E_2(\mathcal{G}) = \{[a_i, a_{i+1}] \times [b_j, b_{j+1}] \subset \mathbb{R}^2 \mid 0 \le i < n, 0 \le j < m\} \cup \{\mathcal{O}\}$, its set of *faces or 2-cells*, where the 'outside face' $\mathcal{O}$ is the face that is exterior to the grating, that is, $\mathcal{O} = \mathbb{R}^2 - \operatorname{int} S$.

Including the 'outside face' $\mathcal{O}$ simplifies the statement of later results.

To emphasis the difference between the combinatorics and the topology, we introduce the **locus** of an i-cell, denoted $|u|$ for $u \in E_i(\mathcal{G})$, defined to be the subset of $\mathbb{R}^2$ that underlies u. For example, if $PQ \in E_1(\mathcal{G})$, then $|PQ| = \{(1-t)P + tQ \mid t \in [0,1]\} \subset \mathbb{R}^2$ when $P = (a_i, b_j)$ and $Q = (a_{i+1}, b_j)$ or $P = (a_i, b_j)$ and $Q = (a_i, b_{j+1})$. Define the following subspaces of $\mathbb{R}^2$:

$$sk_0(\mathcal{G}) = \bigcup\nolimits_{u \in E_0} |u| = E_0(\mathcal{G});$$

$$sk_1(\mathcal{G}) = \bigcup\nolimits_{u \in E_1} |u|; \text{ and}$$

$$sk_2(\mathcal{G}) = \bigcup\nolimits_{u \in E_2} |u| = \mathbb{R}^2.$$

The subspace $sk_0(\mathcal{G})$ is a discrete set and $sk_1(\mathcal{G})$ is a union of line segments. For topological constructions with vertices or edges, such as finding boundaries or interiors, we restrict to these subspaces of $\mathbb{R}^2$.

Suppose we have two elements $u, v \in E_1(\mathcal{G})$ with $u = PQ$ and $v = QR$. The boundaries in the subspace $sk_1(\mathcal{G})$ of $|u|$ and $|v|$ are given by $\mathrm{bdy}_{sk_1} |u| = \mathrm{bdy}_{sk_1} PQ = \{P, Q\}$ and $\mathrm{bdy}_{sk_1} |v| = \{Q, R\}$. The union $|u| \cup |v| = PQ \cup QR$ has boundary $\{P, R\}$, because Q has become an interior point in the subspace topology on $sk_1(\mathcal{G})$, as in the following picture:

$$P \underset{\bullet}{\overset{Q}{\rule{4cm}{0.4pt}}} R$$

We can encode this topological fact in an algebraic manner by associating a union to an addition of cells and boundary to a linear mapping between sums.

Definition 9.1. Let $\mathbb{F}_2$ denote the field with two elements, that is, $\mathbb{F}_2 = \mathbb{Z}/2\mathbb{Z}$. Let the (vector) **space of i-chains** on $\mathcal{G}$ be defined by $C_i(\mathcal{G}) = \mathbb{F}_2[E_i(\mathcal{G})]$, the vector space over $\mathbb{F}_2$ with basis the set $E_i(\mathcal{G})$ for $i = 0, 1, 2$. The **boundary operator** on chains is the linear transformation $\partial \colon C_i(\mathcal{G}) \to C_{i-1}(\mathcal{G})$, for $i = 1, 2$, defined on the basis by $\partial(u) = \sum_l e_l^{i-1}$, where $\partial(u)$ is the sum of the $(i-1)$-cells in $C_{i-1}(\mathcal{G})$ that make up the boundary of the i-cell u, that is, the sum is over $(i-1)$-cells that satisfy $|e_l^{i-1}| \subset \mathrm{bdy}_{sk_i} |u|$.

For example, the boundary operator on a face $ABCD \in E_2(\mathcal{G})$ is given by

$$\partial(ABCD) = AB + BC + CD + DA.$$

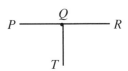

Elements of $C_i(\mathcal{G})$ take the form $\sum_{k=1}^{n} e_k^i$, where $e_k^i \in E_i(\mathcal{G})$ and n is finite. The boundary operator is extended to sums by linearity, $\partial(\sum_{k=1}^{n} e_k^i) = \sum_{k=1}^{n} \partial(e_k^i) \in C_{i-1}(\mathcal{G})$.

The manner in which the combinatorial structure mirrors the topological situation is evident when we compare the formulas:

$$\partial(PQ + QR) = P + Q + Q + R = P + 2Q + R = P + R;$$
$$\mathrm{bdy}_{sk_1} |PQ| \cup |QR| = \{P, R\}.$$

Because $2 = 0$ in $\mathbb{F}_2$, we can drop the term $2Q$. One must be cautious in using these parallel notions—for example,

$$P \overline{} \overset{Q}{\underset{\displaystyle T|}{\vert}} \overline{} R$$

$\partial(PQ + QR + QT) = P + Q + R + T$; while $\mathrm{bdy}_{sk_1} |PQ| \cup |QR| \cup |QT| = \{P, R, T\}$. To compare chains and their underlying sets, we extend the notion of locus to chains. If $c = \sum_{l=1}^{n} e_l^i$, then the *locus of* c is

$$|c| = \left| \sum_{l=1}^{n} e_l^i \right| = \bigcup_{l=1}^{n} |e_l^i|.$$

The addition of chains is related to their locus by a straightforward topological condition.

Lemma 9.2. *If c_1 and c_2 are i-chains, then $|c_1 + c_2| \subset |c_1| \cup |c_2|$, and $|c_1 + c_2| = |c_1| \cup |c_2|$ if and only if* $\mathrm{int}_{sk_i} |c_1| \cap \mathrm{int}_{sk_i} |c_2| = \emptyset$.

Proof. Since the locus of an i-cell e^i is a subset of $|c_j|$ ($j = 1, 2$) whenever the cell occurs in the sum c_j, the union $|c_1| \cup |c_2|$ contains the locus of every cell that appears in either c_1 or c_2. It is possible for a cell to vanish from the algebraic sum if it occurs once in both chains. Thus $|c_1 + c_2| \subset |c_1| \cup |c_2|$. For equality, we need that no i-cell in the sum c_1 appears in c_2. The topological condition on the interiors of cells is equivalent to this condition.   □

Another relation between the combinatorial and the topological holds for 2-chains.

Proposition 9.3. *If $w \in C_2(\mathcal{G})$, then* bdy $|w| = |\partial(w)|$.

Proof. Observe that every edge in $E_1(\mathcal{G})$ is contained in two faces (for this, you need the outside face $\mathcal{O}$ counted among faces). So, if PQ is an edge in $\partial(w)$, then PQ appears only once among the boundaries of faces in w. If x is any point of $|PQ|$, then any open ball centered at x meets the interior of the face w and the exterior of the set $|w|$ and so x is in bdy $|w|$. Conversely, if $x \in$ bdy $|w|$, then x is an element of the locus $|PQ|$ which is an edge PQ in the boundary of a face e^2 in the sum determined by w. Since any open ball centered at x meets points outside $|w|$, the face sharing PQ with e^2 is not in w and so PQ is an edge in $\partial(w)$.   □

A grating can be *refined* by adding vertical and horizontal lines. We could also expand the square region, adding cells that extend the given grating.

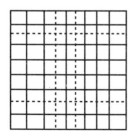
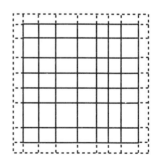

We leave it to the reader to give an expression for the partition of the square region that determines a refinement from the data for a grating. By adding lines we can subdivide the rectangles to have any chosen maximum diameter, no matter how small. We use such an approximation procedure to avoid certain subsets of the plane.

Lemma 9.4. *Let K_1 and K_2 be disjoint compact subsets of $\mathbb{R}^2$ and S a square region with $K_1 \cup K_2 \subset S$. Then any grating $\mathcal{G}$ of S can be refined to a grating $\mathcal{G}^*$ so that no cell of $\mathcal{G}^*$ meets both K_1 and K_2.*

Proof. Since K_1 and K_2 are disjoint and compact, there is a distance $\epsilon > 0$ such that, for any $x \in K_1$ and $y \in K_2$, $d(x, y) \geq \epsilon$. Given the grating $\mathcal{G}$, subdivide the square further so that the diameter of any rectangle is less than $\epsilon/2$. If the locus of a cell contains points x and y, then $d(x, y) < \epsilon/2$ and so it cannot be that $x \in K_1$ and $y \in K_2$. $\square$

We next consider how the combinatorial data are affected by refinement. Of course, certain vertices will be added, edges subdivided and added, and faces subdivided. If $\mathcal{G}$ is refined to a grating $\mathcal{G}^*$ and $c \in C_i(\mathcal{G})$, then we write $c^* \in C_i(\mathcal{G}^*)$ for the i-chain consisting of the the i-cells involved in the subdivision of the i-cells in c. For example, if a 2-cell $ABCD$ is refined by adding a horizontal and a vertical line, then AB is subdivided as AMB, BC as BNC, CD as $CM'D$, and DA as $DN'A$, and we add the vertex P where MM' meets NN'. Then $c^* = AMPN' + MBNP + NCM'P + M'DN'P$. Refinement does not change the locus of an i-cell, that is, $|c| = |c^*|$.

Lemma 9.5. *If c_1 and c_2 are i-chains in $C_i(\mathcal{G})$, then $(c_1 + c_2)^* = c_1^* + c_2^*$ and $(\partial c_1)^* = \partial(c_1^*)$.*

Proof. When we subdivide an i-cell, the number of times (once or not at all) it appears in an i-chain is the same for the parts that constitute its subdivision. Thus the number of times the i-cell appears in the sum will be the same as the number of times the parts appear in the sum of the refined chains and $(c_1 + c_2)^* = c_1^* + c_2^*$.

As for the boundary operator, for 1-chains, subdivision introduces a new intermediate vertex, shared by the 1-cells of the subdivided edge. Thus the new vertices do not appear in $\partial(c^*)$; since refinement does not affect the 0-cells of $\mathcal{G}$, we have $\partial(c^*) = \partial(c) = (\partial c)^*$. For 2-chains,

$$|\partial(c_1^*)| = \text{bdy} \, |c_1^*| = \text{bdy} \, |c_1| = |\partial(c_1)| = |(\partial c_1)^*|.$$

Since $\partial(c_1^*)$ and $(\partial c_1)^*$ are 1-chains in $\mathcal{G}^*$ with the same loci, they are the same 1-chains. □

The combinatorial data provided by chains can be used to study connectedness of subsets of $\mathbb{R}^2$.

Definition 9.6. The **components** of an i-chain $c \in C_i(\mathcal{G})$ are the components of its locus, $|c| \subset \mathbb{R}^2$. We say that two vertices P and Q in a grating $\mathcal{G}$ can be **connected** if there is a 1-chain $\lambda \in C_1(\mathcal{G})$ with $\partial(\lambda) = P + Q$. A subset $A \subset \mathbb{R}^2$ **separates** the vertices P and Q in $\mathbb{R}^2 - A$ if any 1-chain λ connecting P to Q meets A (that is, $|\lambda| \cap A \neq \emptyset$).

We investigate how these combinatorial notions of component and connectedness compare with the usual topological notions.

Proposition 9.7. *Suppose $\mathcal{G}$ is a grating. If c is an i-chain and $c = c_1 + \cdots + c_n$, where each c_j is a maximally connected chain, then the components of $|c|$ are the loci $|c_j|$.*

Proof. If c_j is a maximally connected chain in c, then its locus is connected and $|c_j|$ does not meet the loci of the other chains c_m, $j \neq m$, because if $|c_j| \cap |c_m| \neq \emptyset$, then the chains share an edge ($i = 2$) or a vertex ($i = 1, 2$). In this case, $|c_j| \cup |c_m|$ is connected and $c_j + c_m$ is a connected chain larger than c_j or c_m and hence they are not maximal, a contradiction. Thus the components of c are the maximally connected chains in the sum determined by the chain c. □

Proposition 9.8. *If $A \subset \mathbb{R}^2$ is compact and P and Q are points in $\mathbb{R}^2 - A$, then there is a path in $\mathbb{R}^2 - A$ connecting P to Q if and only if there is a grating $\mathcal{G}$ for which P and Q are vertices, and there is a 1-chain λ with $P + Q = \partial(\lambda)$.*

Proof. Suppose we are given a grating $\mathcal{G}$. If ω is a 1-chain, then we first show that the boundary $\partial(\omega)$ has an even number of vertices. We prove this by induction on the number of 1-cells in the 1-chain. If ω has only one 1-cell, then $\omega = PQ$ and $\partial(\omega) = P + Q$, two vertices. Suppose $\omega = \sum_{i=1}^{n} e_i^1$. Then $\partial(\omega) = \partial(e_1^1) + \sum_{i=2}^{n} \partial(e_i^1)$. By induction, $\sum_{i=2}^{n} \partial(e_i^1)$ is a sum of an even number of vertices. We add to this sum $\partial(e_1^1)$ which consists of two vertices. If either vertex appears in $\sum_{i=2}^{n} \partial(e_i^1)$, then the pair cancels and parity is preserved. Thus $\partial(\omega)$ is a sum of an even number of vertices.

Suppose that $P + Q = \partial(\lambda)$ for some 1-chain $\lambda \in C_1(\mathcal{G})$. If $\lambda = \lambda_1 + \cdots + \lambda_n$ with each λ_i a maximally connected 1-chain in λ, then $\partial(\lambda) = \partial(\lambda_1) + \cdots + \partial(\lambda_n) = P + Q$. Since P and Q must be part of the sum, we can assume that $P + \text{stuff}_1 = \partial(\lambda_1)$ and $Q + \text{stuff}_2 = \partial(\lambda_n)$. Since all the extra stuff must cancel to give $\partial(\lambda) = P + Q$, any vertex appearing in stuff_1 joins λ_1 to another component and so such components are not maximal. Arguing in this manner, we can join P to Q by a connected 1-chain. One can then parameterize the locus of that 1-chain giving a path joining P to Q.

Finally, suppose P and Q are in $\mathbb{R}^2 - A$, an open set. If we can join P to Q by a continuous mapping in $\mathbb{R}^2 - A$, then the image of that path is compact and so some distance $\epsilon > 0$ away from A. Working in the open balls of radius $\epsilon/2$ around points along the curve joining P to Q, we can substitute the path with a path made up of vertical and horizontal line segments. After finding such a path, we extend the line segments to a grating in which the polygonal path is the locus of a 1-chain λ with $\partial(\lambda) = P + Q$. $\qquad\square$

Since a grating $\mathcal{G} = (S, \{a_i\}, \{b_j\})$ is described by finite sets, we can develop some of the purely combinatorial properties of these sets. In particular, the sets $E_i(\mathcal{G})$ are finite, and so we can form the sum of all i-cells into a special i-chain, the *total i-chain*, denoted

$$\Theta^i = \sum_{e^i \in E_i(\mathcal{G})} e^i.$$

Notice that $\partial\Theta^2 = 0$. This follows from the fact that every edge is contained in exactly two cells.

The classes Θ^i give an algebraic expression for the **complement** of an i-chain c, which is denoted by Cc, and defined to be $Cc = \sum_l e_l^i$, where the sum is over all i-cells e_l^i that do *not* appear in the sum c. This sum is easily recovered by observing

$$Cc = c + \Theta^i.$$

Any i-cell appearing in the sum c is cancelled by itself in Θ^i, leaving only the i-cells that did not appear in c.

It is an immediate consequence of the formulas $Cc = c + \Theta^i$ and $\partial\Theta^2 = 0$ that if a 1-chain λ is the boundary of a 2-chain, then $\lambda = \partial(w) = \partial(Cw)$, and so it is the boundary of two complementary 2-chains. This follows from the algebraic version of the complement

$$\partial(Cw) = \partial(w + \Theta^2) = \partial(w) + \partial(\Theta^2) = \lambda.$$

The complement operation leads to a combinatorial version of the Jordan Curve Theorem.

Definition 9.9. An i-chain $c \in C_i(\mathcal{G})$ is an i-**cycle** if $\partial(c) = 0$.

Theorem 9.10. *Every 1-cycle on a grating $\mathcal{G}$ is the boundary of exactly two 2-chains.*

Proof. First observe that the only 2-cycles are 0 and Θ^2. This follows from Proposition 9.3 that $|\partial(c)| = \text{bdy}\,|c|$ for 2-chains. Any nonzero 2-chain c, with $c \neq \Theta^2$, has a nonempty boundary and so is not a 2-cycle.

We prove the theorem by induction on the number of lines involved in the grating. The minimal grating has only the boundary lines of the square region S as edges. The only 2-cells are $ABCD$ and $\mathcal{O}$, both of which satisfy

$$\partial(\mathcal{O}) = \partial(ABCD) = AB + BC + CD + DA.$$

Furthermore, the only nonzero 1-cycle for this grating is $AB + BC + CD + DA = \Theta^1$, so the theorem holds.

Suppose that the theorem holds for a grating $\mathcal{G}$ and we refine $\mathcal{G}$ to $\mathcal{G}^*$ by adding a single vertical line ℓ. (The argument for adding a single horizontal line is analogous.) Suppose that z is a 1-cycle in $C_1(\mathcal{G}^*)$. Define c_ℓ to be the 2-chain which is the sum of all 2-cells with right edges that are on ℓ *and* in the sum z.

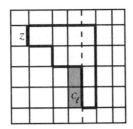

By cancellation, $z + \partial(c_\ell)$ has no edges on ℓ and so we can consider $z + \partial(c_\ell)$ as a 1-chain on $\mathcal{G}$. Furthermore, $\partial(z + \partial(c_\ell)) = \partial(z) + \partial\partial(c_\ell) = 0$, so $z + \partial(c_\ell)$ is a 1-cycle. Since the theorem holds for $\mathcal{G}$, $z + \partial(c_\ell) = \partial(c)$ for some $c \in C_2(\mathcal{G})$. The 2-chain $c^* + c_\ell \in C_2(\mathcal{G})$ has boundary given by

$$\partial(c^* + c_\ell) = (z + \partial(c_\ell))^* + \partial(c_\ell) = (z + \partial(c_\ell)) + \partial(c_\ell) = z.$$

Thus z is the boundary of $c^* + c_\ell$. It is also the boundary of the complement of this 2-chain, $C(c^* + c_\ell) = c^* + c_\ell + \Theta^2 \in C_2(\mathcal{G}^*)$. This follows from $\partial\Theta^2 = 0$.

Finally, we check that no other 2-cell has z as boundary. Suppose $b, b' \in C_2(\mathcal{G})$. If $\partial(b) = \partial(b')$, then $\partial(b + b') = 0$ and so $b + b' = 0$ or $b + b' = \Theta^2$. Then $b = b'$ or $b = Cb'$. Thus, at most two 2-cells can have z as boundary. $\qquad\square$

On a grating, a 1-cycle that is simple (connected without crossings) is a Jordan curve. Theorem 9.10 is a combinatorial version of the Jordan Curve Theorem. The next theorem uses what we have developed so far to establish a general result about separation. It is the key lemma in the proof of the Separation Theorem for Jordan arcs.

The Alexander Theorem. *Suppose K and L are compact subsets of $\mathbb{R}^2$ and $\mathcal{G}$ is a grating of a square S with $K \cup L \subset S$. If $P + Q = \partial(\lambda_1)$ in $\mathbb{R}^2 - K$ and $P + Q = \partial(\lambda_2)$ in $\mathbb{R}^2 - L$, and if $\lambda_1 + \lambda_2 = \partial(w)$ with $|w| \cap K \cap L = \emptyset$, then P is connected to Q by a path that does not meet $K \cup L$.*

Proof. Since $|w| \cap K \cap L = \emptyset$, the compact sets $|w| \cap K$ and $|w| \cap L$ are disjoint. By Lemma 9.4 there is a refinement $\mathcal{G}^*$ of the grating for which no 2-cell meets both $|w| \cap K$ and $|w| \cap L$. Let $w_K = \sum_i e_i^2 \in C_2(\mathcal{G}^*)$, where the sum is over the set $\{e_i^2 \mid e_i^2$ is a 2-cell in w^* and $|e_i^2| \cap K \neq \emptyset\}$. Define the 1-chain $\lambda_0 = \lambda_2^* + \partial(w_K)$. It follows immediately that

$$\partial(\lambda_0) = \partial(\lambda_2^* + \partial(w_K)) = \partial(\lambda_2^*) = P + Q.$$

We know that λ_2^* does not meet L. Since none of the faces of $\mathcal{G}^*$ meet both $|w| \cap K$ and $|w| \cap L$, w_K does not meet L.

To prove the theorem we show that λ_0 does not meet K. Consider the loci:

$$|\lambda_0| = |\lambda_2^* + \partial(w_K)| = |\lambda_1^* + (\lambda_1^* + \lambda_2^* + \partial(w_K))|$$
$$= |\lambda_1^* + \partial(w^* + w_K)| \subset |\lambda_1| \cup \mathrm{bdy}\, |w^* + w_K|.$$

By assumption, λ_1 does not meet K and so λ_1^* does not meet K. In the sum $w^* + w_K$, any 2-cells of w^* that meet K are cancelled by w_K and so $w^* + w_K$ does not meet K. Therefore, $|\lambda_0| \cap K = \emptyset$. Since λ_0 joins P and Q and does not meet $K \cup L$, the theorem is proved. $\square$

Corollary 9.11. *Suppose Λ is a Jordan arc and $\lambda \colon [0,1] \to \Lambda \subset \mathbb{R}^2$ is a parameterization. Let $L_1 = \lambda([0,1/2])$ and $L_2 = \lambda([1/2,1])$. If P is connected to Q in $\mathbb{R}^2 - L_1$ and in $\mathbb{R}^2 - L_2$, then P is connected to Q in $\mathbb{R}^2 - \Lambda$.*

To prove the corollary, simply choose paths that avoid $\lambda(1/2) = L_1 \cap L_2$.

We deduce immediately that if Λ separates P from Q, then one of L_1 or L_2 separates P from Q. From this observation we can give a proof of the Separation Theorem for Jordan arcs. Suppose a Jordan arc Λ separates P from Q; then one of the subsets L_1 or L_2 separates P from Q. Say it is L_1. Then $L_1 = \lambda([0,1/4]) \cup \lambda([1/4,1/2])$ and one of these subsets must separate P from Q by Corollary 9.11. We write L_{1i_2} for a choice of subset that separates P from Q. Halving the relevant subset of $[0,1/2]$ again we can write $L_{1i_2} = L_{1i_21} \cup L_{1i_22}$ and one of these subsets must separate P from Q. Continuing in this manner we get a sequence of nested compact subsets:

$$\cdots \subset L_{1i_2\cdots i_{n-1}i_n} \subset L_{1i_2\cdots i_{n-1}} \subset \cdots \subset L_{1i_2} \subset L_1$$

with the property that each subset separates P from Q. By the inter-section property of nested compact sets (Exercise 6.3), $\bigcap_n L_{1 i_2 \cdots i_n} = R$, a point on Λ. Since the endpoints of the $L_{1 i_2 \cdots i_n}$ constitute a series that converges to R, given an $\epsilon > 0$, there is a natural number N for which $L_{1 i_2 \cdots i_n} \subset B(R, \epsilon)$ for $n \geq N$. By choosing a grating $\mathcal{G}$ to contain P and Q as vertices and for which the subset $B(R, \epsilon) \subset \text{int} \, |w|$ for some $w \in E_2(\mathcal{G})$, we can join P to Q without meeting $L_{1 i_2 \cdots i_N}$, a contradiction. It follows that Λ does not separate P from Q and the theorem is proved.

From this point it is possible to give a proof of the Jordan Curve Theorem using the methods developed so far. Such a proof is outlined in the exercises (or see [63]). We instead use the fundamental group to introduce an integer-valued index whose properties lead to a proof of the Jordan Curve Theorem.

The index of a point not on a Jordan curve

Suppose that $\Omega \in \mathbb{R}^2 - \mathcal{C}$ is a point in $\mathbb{R}^2$ not on a Jordan curve $\mathcal{C}$. To the choice of Ω and a parameterization of $\mathcal{C}$, $\alpha \colon [0,1] \to \mathcal{C} \subset \mathbb{R}^2$, we associate

$$\text{ind}_\Omega(\alpha) = [\alpha] \in \pi_1(\mathbb{R}^2 - \{\Omega\}, \alpha(0)),$$

that is, $\text{ind}_\Omega(\alpha)$ is the homotopy class of the closed curve α in the fundamental group of $\mathbb{R}^2 - \{\Omega\}$ based at $\alpha(0)$. Since the plane with a point removed has the homotopy type of a circle, $\text{ind}_\Omega(\alpha)$ determines an integer via a choice of an isomorphism $\pi_1(\mathbb{R}^2 - \{\Omega\}, \alpha(0)) \cong \mathbb{Z}$. The integer is determined up to a choice of sign and so we write $\text{ind}_\Omega(\alpha) = \pm k \in \mathbb{Z}$ when convenient. We call the choice of integer $\text{ind}_\Omega(\alpha)$ the *index of Ω with respect to α*.

Example. Suppose $\triangle ABC$ is a triangle in the plane and Ω is an interior point. Since $\triangle ABC \simeq S^1$ and Ω may be chosen as a center of S^1, $\text{ind}_\Omega(\triangle ABC) = \pm 1$.

We develop the properties of the index from the basic properties of the fundamental group (Chapters 7 and 8).

Lemma 9.12. *If ℓ is a line in the plane that does not meet $\mathcal{C}$, and Ω and $\mathcal{C}$ lie on opposite sides of ℓ, then $\mathrm{ind}_\Omega(\alpha) = 0$ for any parameterization of $\mathcal{C}$.*

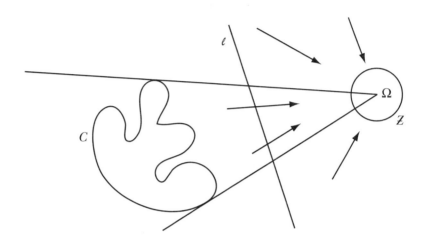

Proof. Let $\mathcal{Z}$ be a small circle centered at Ω entirely in the half-plane determined by ℓ and Ω. We can take $\mathcal{Z}$ as the copy of S^1 which generates $\pi_1(\mathbb{R}^2 - \{\Omega\})$. Since $\mathcal{C}$ is compact and lies on the side of ℓ opposite Ω, all of $\mathcal{C}$ lies in an angle with vertex Ω that is less than two right angles. In the deformation retraction of $\mathbb{R}^2 - \{\Omega\}$ to $\mathcal{Z}$, $\mathcal{C}$ will be taken to a part of $\mathcal{Z}$ where it can be deformed to a point. Thus $\mathrm{ind}_\Omega(\alpha) = 0$ for any choice of parameterization of $\mathcal{C}$. $\square$

The next result takes its name from the shape of the Greek letter θ. Suppose that $\mathcal{C}$ is parameterized in two parts as $\alpha * \gamma \colon [0,1] \to \mathcal{C} \subset \mathbb{R}^2$, where $\alpha(t)$ parameterizes part of the curve, and then $\gamma(t)$ takes over to end at $\gamma(1) = \alpha(0)$. Recall that

$$\alpha * \gamma(t) = \begin{cases} \alpha(2t), & 0 \le t \le 1/2, \\ \gamma(2t - 1), & 1/2 \le t \le 1. \end{cases}$$

Suppose that there is a Jordan arc, parameterized by $\beta \colon [0,1] \to \mathbb{R}^2$, joining $\alpha(1) = \beta(0)$ to $\alpha(0) = \beta(1)$, for which $\beta(t) \notin \mathcal{C}$ for $0 < t < 1$. Then we have three loops beginning at $\alpha(0)$, namely,

$$\omega_0 = \alpha * \gamma, \quad \omega_1 = \alpha * \beta, \quad \text{and} \quad \omega_2 = \beta^{-1} * \gamma,$$

where $\beta^{-1}(t) = \beta(1-t)$. The index of a point Ω that does not lie on $\mathcal{C}$ or on β can be computed for ω_0, ω_1, and ω_2. The next result relates these values.

The Theta Lemma. $\operatorname{ind}_\Omega(\omega_0) = \operatorname{ind}_\Omega(\omega_1) + \operatorname{ind}_\Omega(\omega_2)$ *in* $\pi_1(\mathbb{R}^2 - \{\Omega\}, \alpha(0))$.

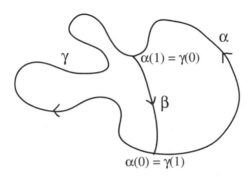

Proof. The binary operation on $\pi_1(\mathbb{R}^2 - \{\Omega\}, \alpha(0))$ is path composition, $*$, which we write as $+$ since $\pi_1(\mathbb{R}^2 - \{\Omega\}, \alpha(0)) \cong \mathbb{Z}$. The lemma follows from the fact that $\beta * \beta^{-1} \simeq c_{\alpha(0)}$, the constant loop at $\alpha(0)$, which is the identity element in the fundamental group:

$$\operatorname{ind}_\Omega(\omega_0) = \operatorname{ind}_\Omega(\alpha * \gamma) = [\alpha * \gamma] = [\alpha * \beta * \beta^{-1} * \gamma]$$
$$= [\alpha * \beta] + [\beta^{-1} * \gamma] = \operatorname{ind}_\Omega(\omega_1) + \operatorname{ind}_\Omega(\omega_2). \qquad \square$$

The next property of the index is crucial to the proof of the Jordan Curve Theorem.

Lemma 9.13. *If Ω and Ω' lie in the same path component of $\mathbb{R}^2 - \mathcal{C}$, then $\operatorname{ind}_\Omega(\alpha) = \operatorname{ind}_{\Omega'}(\alpha)$ for any parameterization of $\mathcal{C}$.*

Proof. Suppose $\lambda\colon [0,1] \to \mathbb{R}^2 - \mathcal{C}$ is a piecewise linear curve joining $\Omega = \lambda(0)$ to $\Omega' = \lambda(1)$. Because $\mathbb{R}^2$ is locally path-connected and $\mathbb{R}^2 - \mathcal{C}$ is an open set, if Ω and Ω' are in the same path component, then it is possible to join them by a piecewise linear curve. We first assume that λ is, in fact, the line segment $\Omega\Omega'$. In the general case, λ will be a finite sequence of line segments connected at endpoints. An induction on the number of such segments completes the argument.

Since the line segment determined by $\Omega\Omega'$ and $\mathcal{C}$ are compact, there is some distance $\epsilon > 0$ between the sets and using this distance we can find a closed rectangle around $\Omega\Omega'$ with the line segment in the center and which is homeomorphic to $[\epsilon, 1 + \epsilon] \times [-\epsilon, \epsilon]$. We use this closed rectangle, contained in $\mathbb{R}^2 - \mathcal{C}$, to construct a homeomorphism $F \colon \mathbb{R}^2 - \{\Omega\} \to \mathbb{R}^2 - \{\Omega'\}$ that leaves $\mathcal{C}$ fixed and so induces an isomorphism

$$F_* \colon \pi_1(\mathbb{R}^2 - \{\Omega\}, \alpha(0)) \longrightarrow \pi_1(\mathbb{R}^2 - \{\Omega'\}, \alpha(0))$$

that sends $[\alpha] \mapsto [\alpha]$. We construct the homeomorphism on the rectangle by first fixing a nice orientation preserving homeomorphism of $[\epsilon, 1 + \epsilon] \times [-\epsilon, \epsilon]$ to the rectangle that takes $[0, 1] \times \{0\}$ to $\Omega\Omega'$. Then make the desired homeomorphism on $[\epsilon, 1 + \epsilon] \times [-\epsilon, \epsilon]$. It is easier to picture the stretching map that will take 0 to 1.

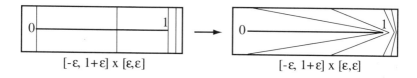

$$[\text{-}\varepsilon,\, 1\text{+}\varepsilon] \times [\varepsilon,\varepsilon] \qquad\qquad [\text{-}\varepsilon,\, 1\text{+}\varepsilon] \times [\varepsilon,\varepsilon]$$

The second parameter, $r \in [-\epsilon, \epsilon]$, is a scaling factor and along each horizontal line segment $[-\epsilon, 1 + \epsilon] \times \{r\}$ we stretch toward the right, pushing $[-\epsilon, 0]$ onto $[\,\epsilon, 1 - \frac{|r|}{\epsilon}]$ and $[0, 1 + \epsilon]$ onto $[1 - \frac{|r|}{\epsilon}, 1 + \epsilon]$. The stretch is the identity along the boundary of the rectangle. The graph of the stretch for various r is shown here:

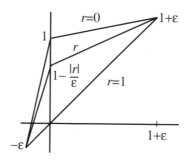

Pasting this change, suitably scaled and rotated, into $\mathbb{R}^2 - \mathcal{C}$ is possible because the stretch is the identity at the boundary. So we can cut out the first closed rectangle and sew in the stretched one to get the desired homeomorphism.

Finally, orienting the boundary of the closed rectangle, we can take its homotopy class as the loop that generates the fundamental group of both spaces $\mathbb{R}^2 - \{\Omega\}$ and $\mathbb{R}^2 - \{\Omega'\}$. Thus, the induced isomorphism F_* takes $[\alpha]$ to $[\alpha]$ and so $\mathrm{ind}_\Omega(\alpha) = \mathrm{ind}_{\Omega'}(\alpha)$ via the isomorphism. $\qquad\square$

The constancy of index along a path and the Theta Lemma have the following important consequence. Suppose that ℓ is a line not passing through $\mathcal{C}$ and Ω is a point in the half-plane determined by ℓ opposite $\mathcal{C}$. Choose points P and Q on the curve such that the line segments $P\Omega$ and $Q\Omega$ do not meet $\mathcal{C}$ except at the endpoints. Parameterize $\mathcal{C}$ by $\alpha\colon [0,1] \to \mathcal{C} \subset \mathbb{R}^2$ with $\alpha(0) = P$ and $\alpha(t_0) = Q$. Let $\gamma_1 = \alpha \circ f_1$, where $f_1\colon [0,1] \to [0,t_0]$ is given by $f_1(s) = t_0 s$. Let $\gamma_2 = \alpha \circ f_2$, where $f_2\colon [0,1] \to [t_0,1]$ is given by $f_2(s) = (1-t_0)s + t_0$. Then $\alpha \simeq \gamma_1 * \gamma_2$. Finally, let $l_1\colon [0,1] \to \mathbb{R}^2$ and $l_2\colon [0,1] \to \mathbb{R}^2$ be the line segments $l_1(t) = (1-t)\alpha(t_0) + t\Omega$ and $l_2(t) = (1-t)\Omega + t\alpha(0)$ for $t \in [0,1]$. These data give the hypotheses for the Theta Lemma with $\omega_0 = \gamma_1 * \gamma_2$, $\omega_1 = \gamma_1 * (l_1 * l_2)$, and $\omega_2 = (l_1 * l_2)^{-1} * \gamma_2$.

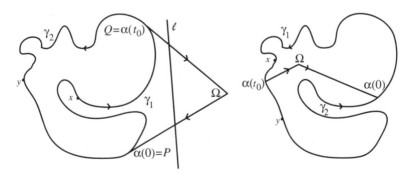

Suppose x lies on γ_1 and y lies on γ_2. Then we can compute the integers $\pm k_1 = \mathrm{ind}_y(\omega_1)$ and $\pm k_2 = \mathrm{ind}_x(\omega_2)$.

Lemma 9.14. *Suppose that R is a point in $\mathbb{R}^2 - (\mathcal{C} \cup P\Omega \cup Q\Omega)$ and suppose that $\mathrm{ind}_R(\omega_1) \neq \mathrm{ind}_y(\omega_1)$ or $\mathrm{ind}_R(\omega_2) \neq \mathrm{ind}_x(\omega_2)$. Then R can be joined to Ω by a path in $\mathbb{R}^2 - \mathcal{C}$.*

Proof. Suppose that $\text{ind}_R(\omega_1) \neq \pm k_1$. Since γ_1 does not separate the plane, there is a path joining R to Ω that does not meet γ_1. Suppose $\zeta: [0,1] \to \mathbb{R}^2$ is such a path with $\zeta(0) = R$ and $\zeta(1) = \Omega$, and $\text{im } \zeta \cap \text{im } \gamma_1 = \emptyset$. Suppose t_1 is the first value in $[0,1]$ with $\zeta(t_1)$ on $l_1 * l_2$, that is, on either line segment $P\Omega$ or ΩQ. Then for $0 \leq t < t_1$, $\text{ind}_{\zeta(t)}(\omega_1)$ is constant. If $\zeta(t)$ meets γ_2 for some $0 \leq t < t_1$, then

$$k_1 \neq \text{ind}_R(\omega_1) = \text{ind}_{\zeta(t)}(\omega_1) = \text{ind}_y(\omega_1) = k_1,$$

a contradiction. Thus ζ on $[0, t_1)$ does not meet γ_1 or γ_2 and so joining ζ restricted to $[0, t_1]$ to the line segment $\zeta(t_1)\Omega$ gives a path from R to Ω. $\qquad\square$

A proof of the Jordan Curve Theorem

To complete a proof of the Jordan Curve Theorem, consider the following subsets of $\mathbb{R}^2 - \mathcal{C}$:

$$U = \{\Omega \in \mathbb{R}^2 - \mathcal{C} \mid \text{ind}_\Omega(\alpha) = 0\}, \quad V = \{R \in \mathbb{R}^2 - \mathcal{C} \mid \text{ind}_R(\alpha) \neq 0\}.$$

For a pair of points $\Omega \in U$ and $R \in V$ there is no path joining them because their indices do not agree. It is clear that $U \neq \emptyset$ because $\mathcal{C}$ is compact and there are lines in the plane that separate the curve from points of index zero. We first prove that $V \neq \emptyset$ and then show that U and V are path-connected.

Let ℓ be a line that does not pass through $\mathcal{C}$. Let Ω lie on the side of ℓ opposite $\mathcal{C}$. Introduce the lines ΩP and ΩQ meeting $\mathcal{C}$ at points $P = \alpha(0)$ and $Q = \alpha(t_0)$, respectively, for some parameterization $\alpha: [0,1] \to \mathcal{C} \subset \mathbb{R}^2$. Introduce the curves $\gamma_1 = \alpha \circ f_1: [0,1] \to \mathbb{R}^2$ with $f_1(s) = t_0 s$, and $\gamma_2 = \alpha \circ f_2: [0,1] \to \mathbb{R}^2$ with $f_2(s) = (1 - t_0)s + t_0$. Thus $\alpha \simeq \gamma_1 * \gamma_2 = \omega_0$. As in the proof of Lemma 9.14, let $l_1(t) = (1 - t)Q + t\Omega$ and $l_2(t) = (1 - t)\Omega + tP$ for $t \in [0,1]$. Form the curve $\omega_1 = \gamma_1 * (l_1 * l_2)$, which travels from $\alpha(0)$ along $\mathcal{C}$ to $\alpha(t_0) = Q$, follows $Q\Omega$ to Ω, then ΩP to $P = \alpha(0)$, and $\omega_2 = (l_1 * l_2)^{-1} * \gamma_2$, which first travels from P along $P\Omega Q$, then follows γ_2 around back to P.

We introduce some other curves in this situation. Let ℓ meet ΩP at R and ΩQ at S. If $l_3(t) = (1 - t)S + tR$, $l_4(t) = (1 - t)Q + tS$, and $l_5(t) = (1 - t)R + tP$, then the curve $\omega_3 = l_5 * \gamma_1 * l_4 * l_3$

together with the triangle $\triangle RS\Omega$ satisfies the conditions for the Theta Lemma. Parameterize the triangle as $l_3^{-1} * l_1' * l_2' = \triangle$, where l_1' is $t \mapsto (1-t)S + t\Omega$ and l_2' is $t \mapsto (1-t)\Omega + tR$. The full curve in the Theta Lemma is $\omega_1' \simeq \triangle * \omega_3$, where ω_1' is ω_1 reparameterized to begin and end at R. Suppose that q is a point in the interior of the triangle $\triangle RS\Omega$. Then we know that $\mathrm{ind}_q(\triangle) = \pm 1$. We apply the Theta Lemma to compute

$$\mathrm{ind}_q(\omega_1) = \mathrm{ind}_q(\omega_1') = \mathrm{ind}_q(\triangle) + \mathrm{ind}_q(\omega_3) = \pm 1 + 0 = \pm 1.$$

We know that $\mathrm{ind}_q(\omega_3) = 0$ since we can separate q from ω_3 by a line parallel to ℓ but close to ℓ.

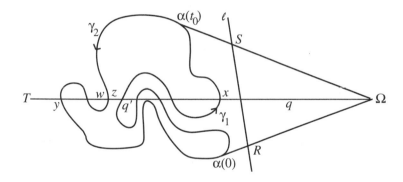

Since $\mathrm{ind}_q(\omega_0) = 0$, we find $\mathrm{ind}_q(\omega_2) = \mp 1$ because $\mathrm{ind}_q(\omega_0) = \mathrm{ind}_q(\omega_1) + \mathrm{ind}_q(\omega_2)$. Extend the ray $\overrightarrow{\Omega q}$ to meet γ_1 first at x, to meet γ_2 last at y. We can compute the indices $\pm k_1 = \mathrm{ind}_y(\omega_1)$ and $\pm k_2 = \mathrm{ind}_x(\omega_2)$ from these points. If T lies on $\overrightarrow{\Omega q}$ far from $\mathcal{C}$, then $\mathrm{ind}_T(\omega_1) = 0$. Since $\overrightarrow{\Omega q}$ meets γ_2 last, by Lemma 9.13, $\mathrm{ind}_T(\omega_1) = \mathrm{ind}_y(\omega_1) = 0 = \pm k_1$. Since $\overrightarrow{\Omega q}$ meets γ_1 first at x, $\mathrm{ind}_x(\omega_2) = \mathrm{ind}_q(\omega_2) = \mp 1 = \pm k_2$.

Suppose $\overrightarrow{\Omega q}$ meets γ_1 last at q' and the next meeting with $\mathcal{C}$ is at w. Let z lie on $\overrightarrow{\Omega q}$ between q' and w. Then $\mathrm{ind}_z(\omega_1) = \mathrm{ind}_w(\omega_1) = \mathrm{ind}_y(\omega_1) = 0$. We also have $\mathrm{ind}_z(\omega_2) = \mathrm{ind}_{q'}(\omega_2) = \mathrm{ind}_x(\omega_2) = \mp 1$. Since $\mathrm{ind}_z(\alpha) = \mathrm{ind}_z(\omega_0) = \mathrm{ind}_z(\omega_1) + \mathrm{ind}_z(\omega_2) = 0 + \mp 1 \neq 0$, we have found $z \in V$ and so $V \neq \emptyset$. Thus $\mathbb{R}^2 - \mathcal{C}$ has at least two components.

We next show that U and V are path-connected. The main tool is
Lemma 9.14. Suppose $\Omega' \in U$, that is, $\Omega' \in \mathbb{R}^2 - C$ and $\mathrm{ind}_{\Omega'}(\alpha) = 0$.
Since $\mathrm{ind}_{\Omega'}(\alpha) = \mathrm{ind}_{\Omega'}(\omega_1) + \mathrm{ind}_{\Omega'}(\omega_2)$ and $\mathrm{ind}_{\Omega'}(\alpha) = 0$, either both
$\mathrm{ind}_{\Omega'}(\omega_i)$ are zero or both are nonzero. In both cases, the values do
not agree with $k_1 = 0$ and $k_2 = \mp 1$. By Lemma 9.14, there is a path
joining Ω' to Ω and so U is path-connected.

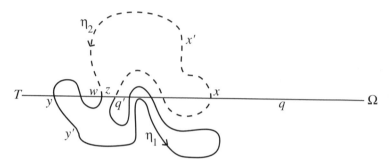

Suppose that M is a point in V. We have shown that the point
z constructed from the intersection of the ray $\overrightarrow{\Omega q}$ with C is also in
V. It suffices to show that there is a path joining M to z. We
apply Lemma 9.14 again. Reparameterize C, $\beta \colon [0, 1] \to C \subset \mathbb{R}^2$,
with $q' = \beta(0)$ and $w = \beta(t_0)$. Let $\eta_1 = \beta \circ f_1$ and $\eta_2 = \beta \circ f_2$
with f_1 and f_2 as before. The curve C is now parameterized with
$\beta \simeq \eta_1 * \eta_2$. Also $\mathrm{ind}_M(\beta) = \mathrm{ind}_M(\alpha) \neq 0$. Let $L_1(t) = (1 - t)w + tz$,
$L_2(t) = (1 - t)z + tq'$. Then

$$\eta_1 * \eta_2 \simeq (\eta_1 * L_1 * L_2) * (L_2^{-1} * L_1^{-1} * \eta_2).$$

Let $\eta_1 * L_1 * L_2 = \overline{\omega}_1$ and $L_2^{-1} * L_1^{-1} * \eta_2 = \overline{\omega}_2$. Take x' on η_1,
y' on η_2, not lying on the line ΩT. Since Ω and T are far from
the curves, $\mathrm{ind}_\Omega(\overline{\omega}_i) = 0 = \mathrm{ind}_T(\overline{\omega}_i)$. Recall that x and q' were on
γ_1, the same parameter range of C, and so $x \in \eta_1$. It follows that
$\pm k_2 = \mathrm{ind}_{x'}(\overline{\omega}_2) = \mathrm{ind}_x(\overline{\omega}_2) = \mathrm{ind}_\Omega(\overline{\omega}_2) = 0$. Similarly, $\pm k_1 = $
$\mathrm{ind}_{y'}(\overline{\omega}_1) = \mathrm{ind}_T(\overline{\omega}_1) = 0$.

We can now apply Lemma 9.14, this time with $k_1 = k_2 = 0$.
Since $\mathrm{ind}_M(\alpha) \neq 0$, there is a path joining M to z. Thus V is path-
connected and we have proved the Jordan Curve Theorem. $\square$

Although we have developed some sophisticated notions to prove
so intuitively simple an assertion, the proof has the virtues of being

rigorous and that it features some ideas that we can develop, namely, the combinatorial and algebraic object given by a grating and the association of an integer or group-valued index to topological objects with nice properties. In the following chapters these ideas take center stage.

Exercises

1. Suppose that X and Y are points in $\mathbb{R}^2$ and $\mathcal{G}$ is a grating with X and Y lying in the interior of two faces in $\mathcal{G}$. A 1-cycle λ is *nonbounding* if any 2-chain w with $\partial(w) = \lambda$ must contain one of the faces containing X or Y. Show that the sum of two nonbounding 1-cycles is not nonbounding.

2. Using the previous exercise, prove that a Jordan curve separates the plane into at most two components. (Hint: Suppose x, y, and z are vertices of a grating $\mathcal{G}$ that contains $\mathcal{C}$. Split the curve into two parts, $\mathcal{C} = \alpha([0, 1/2]) \cup \alpha([1/2, 1])$, that do not separate the points and join them by 1-chains. The subsequent sums are 1-cycles that are nonbounding in the complement of $\{\alpha(0), \alpha(1/2)\}$.)

3. Prove that $\mathbb{R}^2 - \mathcal{C}$ has at least two components using exercise 1.

4. Give an alternate proof of the Separation Theorem for Jordan arcs along the following lines: If Λ is parameterized by $\lambda \colon [0, 1] \to \Lambda \subset \mathbb{R}^2$, then consider the subset $\mathcal{R} = \{r \in [0, 1] \mid \lambda([0, r])$ does not separate the plane$\}$. Show that $\mathcal{R}$ is nonempty, open, and closed.

5. Suppose that $\alpha \colon [0, 1] \to \mathcal{C} \subset \mathbb{R}^2$ and $\beta \colon [0, 1] \to \mathcal{C} \subset \mathbb{R}^2$ are parameterizations of a Jordan curve $\mathcal{C}$ and Ω is a point in $\mathbb{R}^2 - \mathcal{C}$. Show that $\mathrm{ind}_\Omega(\alpha) = \pm \mathrm{ind}_\Omega(\beta)$. Show by example that the sign can change with the parameterization.

6. Suppose K is a subset of $\mathbb{R}^2$ that is homeomorphic to a figure eight (the one-point union of two circles). Generalize the Jordan Curve Theorem to prove that $\mathbb{R}^2 - K$ has three components.

Chapter 10

Simplicial Complexes

The upshot was that he (Poincaré) introduced an entirely new approach to algebraic topology: the concept of complex and the highly elastic algebra going so naturally with it.

SOLOMON LEFSCHETZ, 1970

The gratings of the previous chapter have two nice features—they provide approximations to compact spaces that can be refined to any degree of necessity, and they enjoy a combinatorial and algebraic calculus. These aspects are greatly extended in this chapter and the next. We replace a grating of a square in the plane with a simplicial complex, a particular sort of topological space defined by combinatorial data. Continuous mappings between simplicial complexes can be defined using the combinatorial data. By refining simplicial complexes, we can approximate arbitrary continuous mappings by these combinatorial ones. Approximations are related by homotopies between mappings, giving the homotopy relation further importance. In the next chapter, we will introduce the algebraic structures associated to the combinatorial data. We begin with the basic building blocks.

Definition 10.1. A set of vectors $S = \{\mathbf{v}_0, \ldots, \mathbf{v}_n\}$ in $\mathbb{R}^N$ for N large is in **general position** if the set $\{\mathbf{v}_0 - \mathbf{v}_n, \mathbf{v}_1 - \mathbf{v}_n, \ldots, \mathbf{v}_{n-1} - \mathbf{v}_n\}$

is linearly independent. A set $S = \{\mathbf{v}_0, \ldots, \mathbf{v}_n\}$ in general position is called an n-**simplex** or a simplex of **dimension** n and it determines a subset of $\mathbb{R}^N$ defined by

$$\Delta^n[S] = \{t_0\mathbf{v}_0 + t_1\mathbf{v}_1 + \cdots + t_n\mathbf{v}_n \in \mathbb{R}^N \mid t_i \geq 0, t_0 + \cdots + t_n = 1\}$$
$$= \text{convex hull}(\{\mathbf{v}_0, \ldots, \mathbf{v}_n\}).$$

If the set $S = \{\mathbf{v}_0, \ldots, \mathbf{v}_n\}$ is not in general position, then we say that the n-simplex determined by S is **degenerate**.

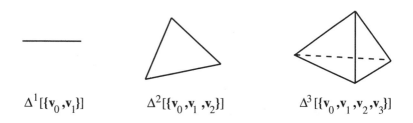

$\Delta^1[\{\mathbf{v}_0, \mathbf{v}_1\}]$ $\qquad$ $\Delta^2[\{\mathbf{v}_0, \mathbf{v}_1, \mathbf{v}_2\}]$ $\qquad$ $\Delta^3[\{\mathbf{v}_0, \mathbf{v}_1, \mathbf{v}_2, \mathbf{v}_3\}]$

For example, a triple $\{\mathbf{v}_0, \mathbf{v}_1, \mathbf{v}_2\}$ is in general position if the points are not collinear. A 0-simplex $\Delta^0[\{\mathbf{v}_0\}]$ is simply the point $\mathbf{v}_0 \in \mathbb{R}^N$. A 1-simplex $\{\mathbf{v}_0, \mathbf{v}_1\}$ determines a line segment $\Delta^1[\{\mathbf{v}_0, \mathbf{v}_1\}]$; $\Delta^2[\{\mathbf{v}_0, \mathbf{v}_1, \mathbf{v}_2\}]$ is a triangle (with its interior) and $\Delta^3[\{\mathbf{v}_0, \mathbf{v}_1, \mathbf{v}_2, \mathbf{v}_3\}]$ is a solid tetrahedron. In general we write $\Delta^n = \Delta^n[S]$ when there is no need to be specific about vertices. When a vertex is repeated, the simplex is degenerate. Degenerate simplices will be important when discussing mappings between simplicial complexes.

In what follows, the combinatorics of sets of vertices play the principal role. We will assume that the vertices determining a simplex are ordered. This assumption is for convenience; in fact, coherent orderings around a simplicial complex determine a useful topological property, orientability (see [**15**], [**29**]), an extra bit of structure to be developed another day.

A point $\mathbf{p} \in \Delta^n$ may be specified uniquely by the coefficients $(t_0, t_1, \ldots, t_n)$. To see this suppose

$$t_0\mathbf{v}_0 + t_1\mathbf{v}_1 + \cdots + t_n\mathbf{v}_n = t_0'\mathbf{v}_0 + t_1'\mathbf{v}_1 + \cdots + t_n'\mathbf{v}_n.$$

Then $(t_0 - t_0')\mathbf{v}_0 + \cdots + (t_n - t_n')\mathbf{v}_n = \mathbf{0}$. Since $\sum_{i=0}^n t_i = \sum_{i=0}^n t_i' = 1$, it follows that $\sum_{i=0}^n (t_i - t_i') = 0$, and so $t_n - t_n' = \sum_{i=0}^{n-1} -(t_i - t_i')$.

In particular,

$$(t_0 - t'_0)\mathbf{v}_0 + \cdots + (t_n - t'_n)\mathbf{v}_n$$
$$= (t_0 - t'_0)(\mathbf{v}_0 - \mathbf{v}_n) + \cdots + (t_{n-1} - t'_{n-1})(\mathbf{v}_{n-1} - \mathbf{v}_n) = \mathbf{0}.$$

Because the set $\{\mathbf{v}_0 - \mathbf{v}_n, \mathbf{v}_1 - \mathbf{v}_n, \ldots, \mathbf{v}_{n-1} - \mathbf{v}_n\}$ is linearly independent, we deduce that $t_i = t'_i$ for all i and so the coefficients are uniquely determined by $\mathbf{p}$. The list of coefficients $(t_0, t_1, \ldots, t_n)$ is called the **barycentric coordinates** of $\mathbf{p} \in \Delta^n$.

Although $\Delta^n[\{\mathbf{v}_0, \ldots, \mathbf{v}_n\}]$ is a subspace of $\mathbb{R}^N$, as a topological space, it is determined by the barycentric coordinates.

Proposition 10.2. *Let* $\boldsymbol{\Delta}^n$ *denote the subspace of* $\mathbb{R}^{n+1}$ *given by* $\boldsymbol{\Delta}^n = \{(t_0, \ldots, t_n) \in \mathbb{R}^{n+1} \mid t_0 + \cdots + t_n = 1, t_i \geq 0\}$. *If* $S = \{\mathbf{v}_0, \ldots, \mathbf{v}_n\}$ *is a set of vectors in general position in* $\mathbb{R}^N$, *then* $\Delta^n[S]$ *is homeomorphic to* $\boldsymbol{\Delta}^n$.

Proof. The mapping $\phi: \boldsymbol{\Delta}^n \to \Delta^n[S]$ given by $\phi(t_0, \ldots, t_n) = t_0\mathbf{v}_0 + \cdots + t_n\mathbf{v}_n$ is a bijection by the uniqueness of barycentric coordinates. The mapping ϕ is given by matrix multiplication and so is continuous. The inverse of ϕ is given by projections on a subspace, and so it too is continuous. $\qquad\square$

The topological properties of $\boldsymbol{\Delta}^n$ are shared with $\Delta^n[S]$ for any other n-simplex. For example, as a subspace of $\mathbb{R}^N$, $\Delta^n[S]$ is compact because $\boldsymbol{\Delta}^n$ is closed and bounded in $\mathbb{R}^{n+1}$.

Proposition 10.3. *The points* $\mathbf{p} \in \Delta^n[S]$ *with barycentric coordinates that satisfy* $t_i > 0$ *for all* i *form an open subset of* $\Delta^n[S]$ *(as a subspace of* $\mathbb{R}^N$*);* $\mathbf{p}$ *is in the boundary of* $\Delta^n[S]$ *if and only if* $t_i = 0$ *for some* i.

Proof. In $\boldsymbol{\Delta}^n \subset \mathbb{R}^{n+1}$, the subset of points with barycentric coordinates $t_i > 0$ is the intersection of the open subsets $U_i = \{(t_0, \ldots, t_n) \in \mathbb{R}^{n+1} \mid t_i > 0\}$ with $\boldsymbol{\Delta}^n$ and so it is an open subset of $\boldsymbol{\Delta}^n$. Its homeomorphic image in $\Delta^n[S]$ is also open in $\Delta^n[S]$.

We can extend the mapping $\phi: \boldsymbol{\Delta}^n \to \Delta^n[S]$ to the subspace Π of $\mathbb{R}^{n+1}$, where

$$\Pi = \{(t_0, \ldots, t_n) \in \mathbb{R}^{n+1} \mid t_0 + \cdots + t_n = 1\},$$

the hyperplane containing $\boldsymbol{\Delta}^n$ in $\mathbb{R}^{n+1}$. The mapping $\widehat{\phi}\colon \Pi \to \mathbb{R}^N$, given by $\widehat{\phi}(t_0, \ldots, t_n) = t_0 \mathbf{v}_0 + \cdots + t_n \mathbf{v}_n$, takes points on the boundary of $\boldsymbol{\Delta}^n$ to points on the boundary of $\Delta^n[S]$. The points on the boundary have some $t_i = 0$ because open sets in $\mathbb{R}^{n+1}$ containing such points must contain points with $t_i < 0$ which map by $\widehat{\phi}$ to points outside $\Delta^n[S]$. Conversely, if a point $\mathbf{p}$ is on the boundary of $\Delta^n[S]$, any open set containing $\mathbf{p}$ meets the complement of $\Delta^n[S]$ and, by a distance argument, points in the image of Π under $\widehat{\phi}$ with negative coordinates. This implies some $t_j = 0$. $\qquad\square$

Notice that a 0-simplex is also its own interior—the topology is discrete on a one-point space. Interesting subsets of a simplex, like the boundary or interior, have nice combinatorial expressions. Define the *face opposite a vertex* $\mathbf{v}_i$ as the subset

$$\partial_i \{\mathbf{v}_0, \ldots, \mathbf{v}_n\} = \{\mathbf{v}_0, \ldots, \widehat{\mathbf{v}_i}, \ldots, \mathbf{v}_n\} = \{\mathbf{v}_0, \ldots, \mathbf{v}_{i-1}, \mathbf{v}_{i+1}, \ldots, \mathbf{v}_n\},$$

where the hat over a vertex means that it is omitted. Any subset of $S = \{\mathbf{v}_0, \ldots, \mathbf{v}_n\}$ determines a *subsimplex* of S, and so a subspace of $\Delta^n[S]$; for example, the subset $T = \{\mathbf{v}_{j_0}, \ldots, \mathbf{v}_{j_k}\}$ determines $\Delta^k[T] = \Delta^k[\{\mathbf{v}_{j_0}, \ldots, \mathbf{v}_{j_k}\}] \subset \Delta^n[S]$. The inclusion is based on the fact that $\sum_i t_{j_i} \mathbf{v}_{j_i} = \sum_{l=0}^n t_l \mathbf{v}_l$, where $t_l = 0$ if $l \neq j_i$.

When $S = \{\mathbf{v}_0, \ldots, \mathbf{v}_n\}$ and $T \subset S$, we denote the inclusion of the subsimplex by $T \prec S$. If $j_0 < j_1 < \cdots < j_k$, then each such face can be obtained by iterating the operation of taking the face opposite some vertex. The combinatorics of the face opposite operators encodes the lower dimensional subsimplices (or faces) of $\Delta^n[S]$. By Proposition 10.3, the geometric boundary of $\Delta^n[S]$ can be expressed combinatorially:

$$\operatorname{bdy} \Delta^n[S] = \Delta^{n-1}[\partial_0 S] \cup \cdots \cup \Delta^{n-1}[\partial_n S] \subset \Delta^n[S].$$

Given any point $\mathbf{p} \in \Delta^n[S]$, writing $\mathbf{p} = t_0 \mathbf{v}_0 + \cdots + t_n \mathbf{v}_n$, we can eliminate the summands with $t_i = 0$ to write $\mathbf{p} = t_{i_0} \mathbf{v}_{i_0} + \cdots + t_{i_m} \mathbf{v}_{i_m}$ with $\sum t_{i_j} = 1$ and $t_{i_j} > 0$ for all j. Thus $\mathbf{p}$ is in the interior of $\Delta^m[\{\mathbf{v}_{i_0}, \ldots, \mathbf{v}_{i_m}\}]$. Because barycentric coordinates are unique, every point in $\Delta^n[S]$ is contained in the interior of a unique subsimplex, $\Delta^m[\{\mathbf{v}_{i_0}, \ldots, \mathbf{v}_{i_m}\}] \subset \Delta^n[S]$.

The simplices $\Delta^n[S]$ form the building blocks of an important class of spaces.

Definition 10.4. A (geometric) **simplicial complex** is a finite collection K of simplices in $\mathbb{R}^N$ satisfying (1) if $S = \{\mathbf{v}_0, \ldots, \mathbf{v}_n\}$ is in K and $T \prec S$ (T is a subset of S), then T is also in K; (2) for S and T in K, if $\Delta^n[S] \cap \Delta^m[T] \neq \emptyset$, then $\Delta^n[S] \cap \Delta^m[T] = \Delta^k[U]$ for some U in K, that is, if simplices of K intersect, then they do so along a common face. The **dimension** of a geometric simplicial complex, $\dim K$, is the largest n for which there is an n-simplex in K.

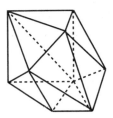 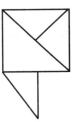

Two collections of triangles in $\mathbb{R}^3$ are shown in the picture. The one on the left represents a simplicial complex, while on the right we have just a union of triangles—this is because the intersections fail to satisfy condition (2) in the definition.

Since n-simplices are homeomorphic to one another for fixed n, it is the collection K of simplices that determines a simplicial complex. We distinguish between the combinatorial data K, collections of sets of vertices, and the topological space determined by the union of the simplices $\Delta^n[S]$ as a subspace of $\mathbb{R}^N$,

$$|K| = \bigcup_{S \in K} \Delta^n[S].$$

The space $|K|$ is called the **realization** of K; $|K|$ is also referred to as the *underlying space* of K [**29**], the *geometric carrier* of K [**15**], or the *polyhedron* determined by K [**35**].

By separating the combinatorial data from the topological data for a simplicial complex, this definition frees us to introduce an abstraction of geometric simplicial complexes.

Definition 10.5. A finite collection of sets

$$L = \{S_\alpha \mid S_\alpha = \{v_{\alpha 0}, \dots, v_{\alpha n_\alpha}\}, 1 \leq \alpha \leq N\}$$

is an **abstract simplicial complex** if whenever $T = \{v_{j_0}, \dots, v_{j_k}\}$ is a subset of S and S is in L, then T is also in L.

In its simplicity there is a gain in flexibility with the notion of an abstract simplicial complex. We can define all sorts of combinatorial objects in this manner (see, for example, [8]). To maintain the connection to topology, we ask if it is possible to associate to every vertex v in an abstract simplicial complex L a point $\mathbf{v}$ in $\mathbb{R}^N$ in such a way that L determines a geometric simplicial complex. The answer is yes, and the proof is an exercise in linear algebra (sketched in the exercises) in which we associate a list of vectors in $\mathbb{R}^N$ in general position to each set S in L. In fact, if the abstract simplicial complex contains a set of cardinality at most $m + 1$, then there is a geometric simplicial complex L' with corresponding sets consisting of vectors in $\mathbb{R}^{2m+1}$ in general position.

Another way to connect with topology is to use the combinatorial data given by an abstract simplicial complex and construct a topological space by gluing simplices together: If $L = \{S \mid S = \{v_0, \dots, v_n\}\}$, then the set of equivalence classes $|L| = [\bigcup_{S \in L} \Delta_S^n]$ associated to the equivalence relation given by $\mathbf{p} \sim \mathbf{q}$ for $\mathbf{p} \in \Delta_S^n$ and $\mathbf{q} \in \Delta_T^m$ if there is a shared face $U \prec S$, $U \prec T$ and $\mathbf{p} = \mathbf{q}$ in $\Delta_U^k \subset \Delta_S^n$ and $\Delta_U^k \subset \Delta_T^m$, that is, we glue the simplices S and T along their shared subsimplex U. We give this space the quotient topology as a quotient of the disjoint union of the simplices Δ_S^n. The reader should check that this quotient construction determines a space homeomorphic to the realization of a geometric simplicial complex built out of vertices in $\mathbb{R}^N$.

The general class of topological spaces modeled by simplicial complexes is the class of the triangulable spaces.

Definition 10.6. A topological space X is said to be **triangulable** if there is an abstract simplicial complex K and a homeomorphism $f\colon X \to |K|$.

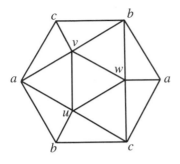
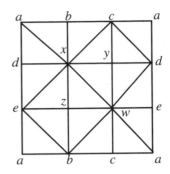

Examples. (1) We can describe triangulable spaces by giving the triangulation explicitly, not as a collection of sets of vectors, but as a collection of simplices with clear gluing data. For example, the diagrams above show how $\mathbb{R}P(2)$ and the torus $S^1 \times S^1$ are triangulable spaces. Notice how the simplices abu and abw in $\mathbb{R}P(2)$ and the simplices abx and abe in the torus share the side ab, encoding the gluing data by the identification of the simplices as shown.

(2) The sphere $S^n \subset \mathbb{R}^{n+1}$ is triangulable in a particularly nice way. Consider the n-simplex $\mathbf{\Delta}^n \subset \mathbb{R}^{n+1}$ for which the vertices are $\mathbf{e}_0, \mathbf{e}_1, \ldots, \mathbf{e}_n$ with $\mathbf{e}_i = (0, 0, \ldots, 0, 1, 0, \ldots, 0)$, where the one is in the $(i+1)$-st place. Consider the point

$$\beta_n = \sum_{i=0}^{n} \frac{1}{n+1} \mathbf{e}_i = (1/(n+1), 1/(n+1), \ldots, 1/(n+1)).$$

This point is the **barycenter** of $\mathbf{\Delta}^n$, and it can be defined for any simplex as the center of gravity of the vertices. We use the barycenter to move the hyperplane in which $\mathbf{\Delta}^n$ lies to pass through the origin. Since $\mathbf{\Delta}^n$ lies in the hyperplane $\Pi = \{(t_0, \ldots, t_n) \mid t_0 + \cdots + t_n = 1\}$, the translated hyperplane through the origin is $\Pi - \beta_n = \{(s_0, \ldots, s_n) \mid s_0 + \cdots + s_n = 0\}$. We identify a copy of S^{n-1} with the intersection of S^n and $\Pi - \beta_n$, that is, elements of $\mathbf{x} \in \mathbb{R}^{n+1}$ satisfying $x_0^2 + x_1^2 + \cdots + x_n^2 = 1$ and $x_0 + x_1 + \cdots + x_n = 0$.

Define the following mapping:

$$\Psi \colon \mathrm{bdy}\, \mathbf{\Delta}^n \to S^{n-1}, \quad \Psi(\mathbf{x}) = \frac{\mathbf{x} - \beta_n}{\|\mathbf{x} - \beta_n\|}.$$

Since the sum of the coordinates of $\mathbf{x}$ is 1, $\mathbf{x} - \beta_n$ lies in $\Pi - \beta_n$ and hence $\Psi(\mathbf{x})$ is in S^{n-1}. Furthermore, Ψ is defined by translation followed by normalization and so Ψ is continuous. Since bdy $\mathbf{\Delta}^n$ is given by $\partial_0 \mathbf{\Delta}^n \cup \cdots \cup \partial_n \mathbf{\Delta}^n$, bdy $\mathbf{\Delta}^n$ is compact. To see that Ψ is a homeomorphism, it suffices, by Proposition 6.9, to show that Ψ has an inverse. Suppose $\mathbf{s} = (s_0, \ldots, s_n)$ is an element of $S^{n-1} = S^n \cap (\Pi - \beta_n)$. Then there is an entry s_k for which $s_k \leq s_i$ for all $0 \leq i \leq n$. Furthermore, since $\sum_i s_i = 0$ and $\sum_i s_i^2 = 1$, we must have $s_k < 0$. Define

$$\Phi \colon S^{n-1} = S^n \cap (\Pi - \beta_n) \to \text{bdy } \mathbf{\Delta}^n, \quad \Phi(\mathbf{s}) = \frac{-1}{s_k(n+1)} \mathbf{s} + \beta_n.$$

To see that $\Phi \circ \Psi$ is the identity, let $\mathbf{x} \in \text{bdy } \mathbf{\Delta}^n$. Then for some $0 \leq k \leq n$, there is an entry $x_k = 0$ in $\mathbf{x}$. It follows that $\mathbf{s} = \Psi(\mathbf{x})$ has entry $s_k = \frac{-1}{(n+1)\|\mathbf{x} - \beta_n\|}$. Furthermore, since $x_i \geq 0$ for all i, s_k is the least entry in $\mathbf{s}$ and so the composite $\Phi \circ \Psi$ gives

$$\Phi \circ \Psi(\mathbf{x}) = \Phi \left(\frac{\mathbf{x} - \beta_n}{\|\mathbf{x} - \beta_n\|} \right)$$

$$= \frac{-1}{(n+1)(-1/((n+1)\|\mathbf{x} - \beta_n\|))} \left(\frac{\mathbf{x} - \beta_n}{\|\mathbf{x} - \beta_n\|} \right) + \beta_n = \mathbf{x}.$$

The opposite composite $\Psi \circ \Phi$ gives the identity on S^{n-1}: because $\|\mathbf{s}\| = 1$ and $s_k < 0$,

$$\Psi \circ \Phi(\mathbf{s}) = \Psi \left(\frac{-1}{(n+1)s_k} \mathbf{s} + \beta_n \right)$$

$$= \frac{(-1/(n+1)s_k)\mathbf{s} + \beta_n - \beta_n}{\|(-1/(n+1)s_k)\mathbf{s} + \beta_n - \beta_n\|} = \mathbf{s}.$$

It follows that bdy $\mathbf{\Delta}^n$ is homeomorphic to S^{n-1}. Since the boundary of $\mathbf{\Delta}^n$ is given as a simplicial complex by the union $\partial_0 \mathbf{\Delta}^n \cup \cdots \cup \partial_n \mathbf{\Delta}^n$, the sphere S^{n-1} is triangulable. This fact will prove useful in Chapter 11.

As with spaces we can apply set-theoretic constructions to simplicial complexes to produce new ones.

Definition 10.7. If K is an abstract simplicial complex and L is a subset of simplices in K, then L is a **subcomplex** of K if whenever $S \prec T$ and $T \in L$, then $S \in L$.

In example (2) above we have shown that $\bigcup_{i=0}^{n} \partial_i \Delta^n = \text{bdy } \Delta^n$ is a subcomplex of Δ^n. In the torus triangulation, notice that the set of simplices $\{dex, xez, xzw, xyw, dyw, dew\}$ together with all the associated subsimplices forms a subcomplex of the torus, whose realization is a cylinder. In the projective plane the subcomplex generated by the collection of 2-simplices $\{abu, auv, uvw, vbw, abw\}$ determines a triangulation of the Möbius band.

Simplicial mappings and barycentric subdivision

How do we compare simplicial complexes? Mappings between simplicial complexes are based on their combinatorial structure.

Definition 10.8. Let K and L be two simplicial complexes. A **simplicial mapping** is a function $\phi \colon K \to L$ satisfying, for all $n \geq 0$, if $S = \{v_0, \ldots, v_n\}$ is an n-simplex in K, then $\{\phi(v_0), \ldots, \phi(v_n)\}$ is a (possibly degenerate) simplex in L. Two simplicial complexes are **isomorphic** if there are simplicial mappings $\phi \colon K \to L$ and $\gamma \colon L \to K$ with $\phi \circ \gamma = \text{id}_L$ and $\gamma \circ \phi = \text{id}_K$. A simplicial mapping $\phi \colon K \to L$ determines a continuous mapping of the associated realizations $|\phi| \colon |K| \to |L|$: If $\phi \colon K \to L$ is a simplicial mapping, then $\mathbf{p} = \sum_{i=0}^{n} t_i \mathbf{v}_i \in |K|$ maps to $|\phi|(\mathbf{p}) = \sum_{i=0}^{n} t_i \phi(\mathbf{v}_i) \in |L|$.

Given a subcomplex $L \subset K$ of a simplicial complex, then the inclusion map $i \colon L \to K$ is a simplicial mapping. Also, a composite of simplicial mappings $K \xrightarrow{\phi} L \xrightarrow{\gamma} M$ is a simplicial mapping.

Since the mapping $|\phi| \colon |K| \to |L|$ associated to a simplicial mapping is linear on each simplex, it is continuous. Notice that there are only finitely many continuous mappings $|K| \to |L|$ that are realized in this manner. Because there are only finitely many 0-simplices in K and L, there are only finitely many vertex-to-vertex mappings, of which the simplicial mappings are a subset. In what follows, we construct more simplicial mappings between $|K|$ and $|L|$. To do so, we refine a simplicial complex in order to make approximations. A refinement of a grating in Chapter 9 was accomplished by the addition of line segments, subdividing the rectangles into smaller cells. To refine a simplicial complex, we subdivide the simplices.

Definition 10.9. Let K be a simplicial complex. The **barycentric subdivision** of K, denoted sd K, is the simplicial complex whose simplices are given by

$$\{\beta(S_0), \beta(S_1), \ldots, \beta(S_r)\}, \text{ where } S_i \in K \text{ and } S_0 \prec S_1 \prec \cdots \prec S_r.$$

Here $\beta(S) = \beta(\{\mathbf{v}_0, \ldots, \mathbf{v}_n\}) = \sum_{i=0}^{n} \frac{1}{n+1} \mathbf{v}_i$ is the barycenter of $\Delta^n[S]$ for S in K. If $\phi \colon K \to L$ is a simplicial mapping, then the barycentric subdivision of ϕ is the simplicial mapping sd $\phi \colon$ sd $K \to$ sd L given on vertices by sd $\phi(\beta(S)) = \beta(\phi(S))$.

The operation $K \mapsto$ sd K may be summarized: First find the barycenters of every simplex in K, then subdivide the simplices of K into new simplices organized by the subset ordering of simplices, $S \prec T$. For example, a one-simplex $\{a, b\}$ is realized by the line segment ab. The barycenter is the midpoint of ab and the barycentric subdivision sd $\{a, b\}$ has two one-simplices $\{a, \beta_1\}$ and $\{\beta_1, b\}$ corresponding to $\{a\} \prec \{a, b\}$ and $\{b\} \prec \{a, b\}$. The barycentric subdivision of a two-simplex, $\Delta^2[\{a, b, c\}]$, has six two-simplices as in the following picture:

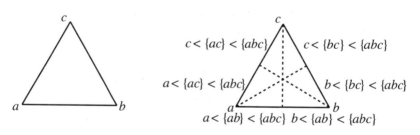

The effect of barycentric subdivision on a simplicial mapping is to send the new barycenters of simplices in K to the corresponding barycenters of the image simplices in L.

To understand the kind of approximation the barycentric subdivision provides, we introduce the *diameter of a simplex*: Let K be a simplicial complex, realized in $\mathbb{R}^N$. Then

$$\text{diam } S = \max\{\|\mathbf{v}_i - \mathbf{v}_j\| \mid i \neq j, \ S = \{\mathbf{v}_0, \ldots, \mathbf{v}_q\}\}.$$

The diameter depends on the embedding of $|K|$ in $\mathbb{R}^N$, but this dependence will not affect the combinatorial use of subdivision.

Proposition 10.10. *If S is a q-simplex in K, a geometric simplicial complex, then for any simplex $T \in$ sd K with $\Delta^p[T] \subset \Delta^q[S]$, we have* diam $T \leq \frac{q}{q+1}$ diam S.

Proof. We proceed by induction on q. If $q = 1$, then $\Delta^1[S]$ is a line segment and the simplices of the barycentric subdivision are halves of the segment with diameter equal to $1/2$ the length of the segment. Assume the result for simplices of dimension less than $q \geq 2$.

A p-simplex $T \in$ sd K can be written as

$$T = \left\{ \mathbf{v}_{\sigma(0)}, \frac{\mathbf{v}_{\sigma(0)} + \mathbf{v}_{\sigma(1)}}{2}, \frac{\mathbf{v}_{\sigma(0)} + \mathbf{v}_{\sigma(1)} + \mathbf{v}_{\sigma(2)}}{3}, \ldots, \right.$$
$$\left. \frac{\mathbf{v}_{\sigma(0)} + \mathbf{v}_{\sigma(1)} + \cdots + \mathbf{v}_{\sigma(p)}}{p+1} \right\},$$

where σ is some permutation of $(0, 1, \ldots, q)$. If $p < q$, then we are done because T is a simplex in the barycentric subdivision of a face of S. When $p = q$, write the vertices of T as $T = \{\mathbf{w}_0, \mathbf{w}_1, \ldots, \mathbf{w}_q\}$. The diameter of T is given by $\|\mathbf{w}_{i_0} - \mathbf{w}_{j_0}\| = \max\{\|\mathbf{w}_i - \mathbf{w}_j\| \mid \mathbf{w}_i, \mathbf{w}_j \in T\}$. If i_0 and j_0 are less than q, then the diameter of T is achieved on the face $\partial_q T$ and we deduce

$$\|\mathbf{w}_{i_0} - \mathbf{w}_{j_0}\| \leq \frac{q-1}{q} \operatorname{diam} \partial_q S \leq \frac{q}{q+1} \operatorname{diam} S.$$

If one of i_0 or j_0 is q, then we first observe the following estimate:

$$\left\| \mathbf{v}_i - \frac{\mathbf{v}_{\sigma(0)} + \mathbf{v}_{\sigma(1)} + \cdots + \mathbf{v}_{\sigma(q)}}{q+1} \right\|$$
$$= \left\| \sum_{j=0}^{q} \frac{1}{q+1}(\mathbf{v}_i - \mathbf{v}_j) \right\| \leq \sum_{j=0}^{q} \frac{1}{q+1} \|\mathbf{v}_i - \mathbf{v}_j\|$$
$$\leq \frac{q}{q+1} \max\{\|\mathbf{v}_i - \mathbf{v}_j\|\} = \frac{q}{q+1} \operatorname{diam} S.$$

This proves the proposition. $\square$

We define a measure of the refinement of a simplicial complex by taking the maximum of the diameters of the constituent simplices, the *mesh* of K,

$$\operatorname{mesh}(K) = \max\{\operatorname{diam} S \mid S \in K\}.$$

Corollary 10.11. *If K has dimension q, then*

$$\text{mesh}(\text{sd } K) \leq \frac{q}{q+1}\text{mesh}(K).$$

By iterating barycentric subdivision, we can make the simplices in $\text{sd}^N K$ as small as we like: For any $\epsilon > 0$, there is an N with $\text{mesh}(\text{sd}^N K) \leq \left(\frac{q}{q+1}\right)^N \text{mesh}(K) < \epsilon$.

How does barycentric subdivision affect the topological space $|K|$?

Theorem 10.12. *If K is a geometric simplicial complex, then $|\text{sd } K| = |K|$.*

Proof ([28]). Suppose $\mathbf{p} \in |K|$. Then we can write $\mathbf{p} = \sum_{i=0}^{q} t_i \mathbf{v}_i \in \Delta^q[S]$ with $S = \{\mathbf{v}_0, \ldots, \mathbf{v}_q\}$. Permute the values $\{t_i\}$ to bring them into descending order:

$$t_{\sigma(0)} \geq t_{\sigma(1)} \geq \cdots \geq t_{\sigma(q)} \geq 0.$$

Next solve the matrix equation

$$\begin{pmatrix} 1 & \frac{1}{2} & \frac{1}{3} & \cdots & \frac{1}{q+1} \\ 0 & \frac{1}{2} & \frac{1}{3} & \cdots & \frac{1}{q+1} \\ 0 & 0 & \frac{1}{3} & \cdots & \frac{1}{q+1} \\ \vdots & \vdots & \vdots & \cdots & \vdots \\ 0 & 0 & 0 & \cdots & \frac{1}{q+1} \end{pmatrix} \begin{pmatrix} s_0 \\ s_1 \\ s_2 \\ \vdots \\ s_q \end{pmatrix} = \begin{pmatrix} t_{\sigma(0)} \\ t_{\sigma(1)} \\ t_{\sigma(2)} \\ \vdots \\ t_{\sigma(q)} \end{pmatrix}.$$

The solution exists and is unique. Furthermore, by solving from the bottom up, the solution satisfies $s_q = (q+1)t_{\sigma(q)}$ and $s_{j-1} = j(t_{\sigma(j-1)} - t_{\sigma(j)}) \geq 0$. Summing the values of s_j we get

$$\sum_{j=0}^{q} s_j = s_0 + 2((1/2)s_1) + 3((1/3)s_2) + \cdots + (q+1)((1/(q+1))s_q)$$

$$= (s_0 + (1/2)s_1 + (1/3)s_2 + \cdots + (1/(q+1))s_q)$$
$$+ ((1/2)s_1 + (1/3)s_2 + \cdots + (1/(q+1))s_q)$$
$$+ \cdots + (1/(q+1))s_q$$
$$= t_{\sigma(0)} + t_{\sigma(1)} + \cdots + t_{\sigma(q)} = t_0 + \cdots + t_q = 1.$$

Thus $(s_0, \ldots, s_q)$ are the barycentric coordinates of $\mathbf{p}$ in the simplex with

$$\mathbf{p} = s_0 \mathbf{v}_{\sigma(0)} + s_1 \left(\frac{\mathbf{v}_{\sigma(0)} + \mathbf{v}_{\sigma(1)}}{2} \right) + s_2 \left(\frac{\mathbf{v}_{\sigma(0)} + \mathbf{v}_{\sigma(1)} + \mathbf{v}_{\sigma(2)}}{3} \right)$$

$$+ \cdots + s_q \left(\frac{\mathbf{v}_{\sigma(0)} + \mathbf{v}_{\sigma(1)} + \cdots + \mathbf{v}_{\sigma(q)}}{q + 1} \right).$$

Thus $\mathbf{p}$ lies in the q-simplex $\Delta^q[T]$, where $T \in \mathrm{sd}\, K$ is given by

$$T = \{ \beta(\{\mathbf{v}_{\sigma(0)}\}), \beta(\{\mathbf{v}_{\sigma(0)}, \mathbf{v}_{\sigma(1)}\}), \beta(\{\mathbf{v}_{\sigma(0)}, \mathbf{v}_{\sigma(1)}, \mathbf{v}_{\sigma(2)}\}), \ldots,$$
$$\beta(\{\mathbf{v}_{\sigma(0)}, \mathbf{v}_{\sigma(1)}, \ldots, \mathbf{v}_{\sigma(q)}\}) \}.$$

This proves that $|K| \subset |\mathrm{sd}\, K|$. The inclusion $|\mathrm{sd}\, K| \subset |K|$ follows by rewriting the expression for a point in the barycentric coordinates of $\mathrm{sd}\, K$ in terms of the contributing vertices of K by rearranging terms. $\square$

Barycentric subdivision leads to a notion of approximation. Given a continuous mapping $f \colon |K| \to |L|$, we seek a simplicial mapping $\phi \colon K \to L$ that *approximates* f in some sense. Since we can replace $|K|$ with $|\mathrm{sd}^n K|$, where $\mathrm{sd}^n K$ denotes the iterated barycentric subdivision of K, $\mathrm{sd}^0 K = K$, and $\mathrm{sd}^n K = \mathrm{sd}(\mathrm{sd}^{n-1} K)$, then we can approximate f by using simplicial mappings between subdivisions of the complexes involved. To make precise what we mean by an approximation, we introduce a point-set notion.

Definition 10.13. If v is a vertex in a simplicial complex K, then the **star** of v, $\mathrm{star}_K(v)$, is the collection of all simplices in K for which v is a vertex. The **open star** of v, $O_K(v)$, is the union of the interiors of simplices in K with v as a vertex,

$$\mathrm{star}_K(v) = \bigcup\nolimits_{\{v\} \prec S} \Delta^n[S], \quad O_K(v) = \bigcup\nolimits_{\{v\} \prec S} \mathrm{int}\, \Delta^n[S].$$

The stars of vertices can be used to recognize simplices in a simplicial complex.

Lemma 10.14. *Suppose $v_0, v_1, \ldots, v_n$ are vertices in a simplicial complex K. Then $\{v_0, \ldots, v_q\}$ is a simplex in K if and only if $\bigcap_{i=0}^q O_K(v_i) \neq \emptyset$. If $\mathbf{p} \in |K|$, then $\mathbf{p} \in O_K(v)$ if and only if $\mathbf{p} = \sum_{i=0}^q t_i v_i$ with $v = v_j$ for some $0 \leq j \leq q$ and $t_j \neq 0$.*

Proof. If $S = \{v_0, \ldots, v_q\}$ is a q-simplex in K, then $\operatorname{int} \Delta^q[S] \subset O_K(v_i)$ for $i = 0, \ldots, q$. Hence $\bigcap_{i=0}^{q} O_K(v_i) \neq \emptyset$.

Suppose $\mathbf{p} \in \bigcap_{i=0}^{q} O_K(v_i) \neq \emptyset$. Then $\mathbf{p} = \sum t_j w_j \in \Delta^r[S]$ with $\{v_0, \ldots, v_q\} \subset \{w_0, \ldots, w_r\}$. Furthermore, if $w_{m_i} = v_i$, then $t_{m_i} > 0$. Thus all of the v_i appear in the barycentric coordinates of $\mathbf{p}$ and so the subset of S, $\{v_0, \ldots, v_q\}$, is a simplex in K. $\qquad\square$

To approximate a continuous mapping $f \colon |K| \to |L|$ by a simplicial mapping $\phi \colon K \to L$, we expect that points in $f(\Delta^q[S])$ are 'close' to points in $|\phi|(\Delta^q[S])$.

Definition 10.15. If K and L are simplicial complexes and $f \colon |K| \to |L|$ is a continuous function, then a simplicial mapping $\phi \colon K \to L$ is a **simplicial approximation** to f if whenever $\mathbf{p} \in |K|$, then $f(\mathbf{p}) \in \Delta^q[T]$ for $T \in L$ implies $|\phi|(\mathbf{p}) \in \Delta^q[T]$.

This definition can be difficult to establish, but there is a more convenient condition for our purposes that works in a manner analogous to the way open sets simplify continuity arguments when compared with the classical ϵ-δ arguments.

Proposition 10.16. *A simplicial mapping $\phi \colon K \to L$ is a simplicial approximation to a continuous mapping $f \colon |K| \to |L|$ if and only if, for any vertex v of K, we have*

$$f(O_K(v)) \subset O_L(\phi(v)),$$

that is, the image of the open star of v under f is contained in the open star of $\phi(v)$, a vertex of L.

Proof. Suppose $\mathbf{p} \in O_K(v)$ for some vertex $v \in K$. Then $\mathbf{p} \in \operatorname{int} \Delta^q[S]$ for some unique $S \in K$ with $v \in S$. Because ϕ is a simplicial mapping, $\phi(S) = T$ for some simplex in L, and $|\phi|(p) \in \operatorname{int} \Delta^{q'}[T'] \subset O_L(\phi(v))$ for some $T' \prec T$. Since ϕ is a simplicial approximation to f, if $\mathbf{p} \in \Delta^r[S']$ for $S \prec S'$ and $f(\mathbf{p}) \in \operatorname{int} \Delta^s[T'']$ for some $T'' \in L$, then $|\phi|(\mathbf{p}) \in \Delta^s[T'']$. Since points lie in unique interiors of simplices, $|\phi|(\mathbf{p}) \in \operatorname{int} \Delta^{q'}[T']$ implies that $T' \prec T''$ and so $\phi(v) \in T''$. Therefore, $f(\mathbf{p}) \in O_L(\phi(v))$.

We introduce a weaker notion than a simplicial mapping. Let $K_0 = \{v \in K \mid \{v\}$, a 0-simplex in $K\}$. A *vertex map* $\phi \colon K_0 \to L_0$

satisfies the condition that if $v \in K$ is a vertex, then $\phi(v) \in L$ is also a vertex. Suppose also, for every vertex $v \in K_0$, that $f(O_K(v)) \subset O_L(\phi(v))$. Suppose that $S \in K$ is a simplex and $S = \{v_0, \ldots, v_q\}$. Then

$$f\left(\bigcap_i O_K(v_i)\right) \subset \bigcap_i f(O_K(v_i)) \subset \bigcap_i O_L(\phi(v_i)).$$

Since int $\Delta^q[S] \subset \bigcap_i O_K(v_i)$, this intersection is nonempty, and $\phi(S) = \{\phi(v_0), \ldots, \phi(v_q)\}$ is a simplex in L. This establishes that a vertex mapping ϕ with $f(O_K(v)) \subset O_L(\phi(v))$, for all v, is a simplicial mapping. Furthermore, if $\mathbf{p} \in$ int $\Delta^q[S]$ and $f(\mathbf{p}) \in$ int $\Delta^r[T]$ for some $T \in L$, then for each vertex v_i of S, $f(\mathbf{p}) \in f(O_K(v_i)) \subset O_L(\phi(v_i))$, and so $\phi(v_i) \in T$. It follows that $\phi(S) \prec T$ and so $|\phi|(\mathbf{p}) \in \Delta^r[T]$. Therefore, ϕ is a simplicial approximation to f. $\quad\square$

Example. In Theorem 10.12 we proved that $|\operatorname{sd} K| = |K|$. Is there a simplicial approximation to the identity mapping? Consider the vertex mapping $\lambda\colon \operatorname{sd} K \to K$, defined by

$$\lambda\colon \beta(S) = \beta(\{v_0, \ldots, v_q\}) \mapsto v_q.$$

To see that we have a simplicial approximation, we check that $O_{\operatorname{sd} K}(\beta(S)) \subset O_K(v_q)$. A simplex with $\beta(S)$ as a vertex takes the form $T = \{\beta(S_0), \beta(S_1), \ldots, \beta(S_n)\}$ with $S_1 \prec S_2 \prec \cdots \prec S_n$ in K and $S = S_j$ for some j. If $\mathbf{p} \in$ int $\Delta^q[T]$, then $\mathbf{p} = \sum_i t_i \beta(S_i)$ with $t_i > 0$. We can rewrite the barycenters as the averages of the vertices in S_i for $i = 0$ to q, and we get $\mathbf{p} = \sum_k u_k w_k$ with $u_k > 0$ and $w_k \in K$ for all k. Since v_q is among the vertices and its barycentric coordinate is positive, $\mathbf{p} \in O_K(v_q)$. Thus λ is a simplicial approximation to id$\colon |\operatorname{sd} K| \to |K|$. In fact, we did not need to choose the last vertex v_q to define λ. As the argument shows, any choice of vertex from S for each $S \in K$ will do. This added flexibility will come in handy later.

The topology of a triangulable space may be used to show that simplicial approximations are plentiful.

Simplicial Approximation Theorem. *Given two simplicial complexes K and L and a continuous mapping $f\colon |K| \to |L|$, then there is a nonnegative integer r and a simplicial mapping $\phi\colon \operatorname{sd}^r K \to L$ with ϕ a simplicial approximation to f.*

Proof. We use the fact that $|K|$ and $|L|$ are compact metric spaces. Suppose $\dim K = n$. The collection $\{f^{-1}(O_L(w)) \mid w \text{ a vertex in } L\}$ is an open cover of $|K|$. By Lebesgue's Lemma (Chapter 6) the cover has a Lebesgue number $\delta_K > 0$. Iterating barycentric subdivision, we can subdivide K until

$$\text{mesh}(\text{sd}^r K) \le \left(\frac{n}{n+1}\right)^r \text{mesh}(K) < \delta_K/2.$$

This is possible because $(\frac{n}{n+1})^r$ goes to zero as r goes to infinity. It follows that $\text{sd}^r K$ has all simplices of diameter less than $\delta_K/2$ and so, for each $v \in \text{sd}^r K$, the diameter of $O_K(v)$ is less than δ_K. Thus each $O_K(v)$ is contained in some $f^{-1}(O_L(w))$. This determines a vertex map $\phi \colon v \mapsto w$, which satisfies $f(O_K(v)) \subset O_L(\phi(v))$, a simplicial approximation. $\qquad\square$

Simplicial approximations exist in abundance. How are these combinatorial mappings related to their approximated topological mappings? What relation is there between two simplicial approximations of the same continuous mapping? We can answer these questions with the homotopy relation between continuous mappings. This relationship formed the basis for the combinatorial nature of some of the earliest developments in topology (see, for example, [**10**]).

Proposition 10.17. *If a simplicial mapping* $\phi \colon K \to L$ *is a simplicial approximation to a continuous mapping* $f \colon |K| \to |L|$, *then* $|\phi|$ *is homotopic to* f.

Proof. Suppose that $\mathbf{p} \in \text{int } \Delta^q[S]$ for $S \in K$ and $S = \{v_0, \dots, v_q\}$. By Lemma 10.14, $\mathbf{p} \in \bigcap_{v_i \in S} O_K(v_i)$. It follows that

$$f(\mathbf{p}) \in \bigcap_{v_i \in S} f(O_K(v_i)) \subset \bigcap_{v_i \in S} O_L(\phi(v_i)).$$

Therefore, $\{\phi(v_0), \dots, \phi(v_q)\}$ is a simplex in L and the convex set $\Delta^q[\phi(S)]$ contains both $|\phi|(\mathbf{p})$ and $f(\mathbf{p})$. We define a homotopy on $\text{int } \Delta^q[S]$ by

$$H(\mathbf{p}, t) = tf(\mathbf{p}) + (1-t)|\phi|(\mathbf{p}).$$

The homotopy extends to all of $|K|$ by Theorem 4.4 and so $f \simeq |\phi|$. $\qquad\square$

It follows from the proposition that two, possibly different, simplicial approximations to a given continuous function have homotopic realizations. The simplicial mappings also enjoy a further combinatorial property.

Definition 10.18. Two simplicial mappings ϕ and $\psi\colon K \to L$ are said to be **contiguous** if, for all simplices $S \in K$, the set $\phi(S) \cup \psi(S)$ is a simplex in L.

Lemma 10.19. *Suppose* $f\colon |K| \to |L|$ *is a continuous function for which* ϕ *and* $\psi\colon K \to L$ *are simplicial approximations to* f. *Then* ϕ *and* ψ *are contiguous.*

Proof. Suppose S is a simplex in K with $S = \{v_0, \ldots, v_q\}$. Then for $\mathbf{p} \in \mathrm{int}\, \Delta^q[S]$, we have

$$f(\mathbf{p}) \in f\left(\bigcap_i O_K(v_i)\right) \subset \bigcap_i f(O_K(v_i)) \subset \bigcap_i O_L(\phi(v_i)) \cap O_L(\psi(v_i)).$$

Since this intersection is not empty, the collection $\phi(S) \cup \psi(S)$ is a simplex in L. □

The condition of being contiguous is combinatorial—we are only checking that unions of images of sets of vertices in K appear among the sets of vertices of L. The following results show that contiguity encodes the relation of homotopy very well.

Proposition 10.20. *Contiguous simplicial mappings have homotopic realizations.*

Proof. If $\mathbf{p} \in \mathrm{int}\, \Delta^q[S] \subset |K|$, then the points $|\phi|(\mathbf{p})$ and $|\psi|(\mathbf{p})$ lie in the simplex of L given by $\phi(S) \cup \psi(S)$. The homotopy $H(\mathbf{p}, t) = (1 - t)|\phi|(\mathbf{p}) + t|\psi|(\mathbf{p})$ is well defined, continuous, and establishes $|\phi| \simeq |\psi|$. □

A partial converse to Proposition 10.20 is the following theorem.

Theorem 10.21. *Suppose that* f *and* g *are continuous mappings* $|K| \to |L|$ *and* f *is homotopic to* g. *Then there exist simplicial mappings* ϕ *and* $\psi\colon \mathrm{sd}^N K \to L$ *with* ϕ *a simplicial approximation to* f, ψ *a simplicial approximation to* g, *and there is a sequence of simplicial mappings* $\phi = \phi_0,\ \phi_1,\ \ldots,\ \phi_{n-1}, \phi_n = \psi$ *with* ϕ_i *contiguous to* ϕ_{i+1} *for* $0 \le i \le n - 1$.

Proof. Let $H\colon |K| \times [0,1] \to |L|$ be a homotopy with $H(\mathbf{p}, 0) = f(\mathbf{p})$ and $H(\mathbf{p}, 1) = g(\mathbf{p})$. Cover $|K| \times [0,1]$ with the open cover $\{H^{-1}(O_L(w)) \mid w \text{ is a vertex of } L\}$. Since $|K| \times [0,1]$ is compact, by a careful use of Lebesgue's Lemma, we can find a partition of $[0,1]$, $0 = t_0 < t_1 < \cdots < t_{n-1} < t_n = 1$, such that, for any $\mathbf{p} \in |K|$, $H(\mathbf{p}, t_{i-1})$ and $H(\mathbf{p}, t_i)$ lie in $O_L(w)$ for some vertex $w \in L$. Define the functions $h_i\colon |K| \to |L|$ by $h_i(\mathbf{p}) = H(\mathbf{p}, t_i)$. Construct another open cover of $|K|$ defined as $\mathcal{U} = \mathcal{U}_1 \cup \cdots \cup \mathcal{U}_n$, where

$$\mathcal{U}_i = \{h_i^{-1}(O_L(w)) \cup h_{i-1}^{-1}(O_L(w)) \mid w \text{ a vertex in } L\}.$$

Subdivide K enough times so that the simplices in $\mathrm{sd}^N K$ are finer than the cover $\mathcal{U}$. Let $\phi_i\colon \mathrm{sd}^N K \to L$ be the vertex mapping which satisfies $h_i(O_K(v)) \cup h_{i-1}(O_K(v)) \subset O_L(\phi_i(v))$ for each vertex $v \in \mathrm{sd}^N K$. By construction, ϕ_i is a simplicial approximation to h_i and h_{i-1}. Regrouping these data, we find that ϕ_i and ϕ_{i+1} are both simplicial approximations to h_i and hence ϕ_i and ϕ_{i+1} are contiguous by Lemma 10.19. Since $h_0 = f$ and $h_n = g$, $\phi = \phi_0$ is a simplicial approximation of f, and $\psi = \phi_n$ is a simplicial approximation to g. This proves the theorem. $\qquad\square$

We close with a consequence of these ideas. Suppose X and Y are triangulable spaces. Then the set of homotopy classes of mappings from X to Y is denoted by $[X, Y]$, as introduced in Chapter 7. We can replace this set by $[|K|, |L|]$, where $|K|$ is homeomorphic to X and $|L|$ is homeomorphic to Y. By the Simplicial Approximation Theorem, for each homotopy class $[f] \in [|K|, |L|]$, there is a simplicial mapping $\phi\colon \mathrm{sd}^r K \to L$ with $[|\phi|] = [f]$. Furthermore, by Proposition 10.20 and Theorem 10.21, different choices of representative for $[f]$ always stay in the same homotopy class of the realization of the simplicial approximation.

Let $\mathcal{S}(K, L)$ denote the set of simplicial mappings from K to L. Because K and L involve only finitely many simplices, $\mathcal{S}(K, L)$ is a finite set. With this notation, the Simplicial Approximation Theorem implies that the mapping

$$\Theta\colon \bigcup_{N \geq 0} \mathcal{S}(\mathrm{sd}^N K, L) \longrightarrow [|K|, |L|], \qquad \Theta(\phi) = [|\phi|],$$

is onto. The union of countably many finite sets is countable and so we have proved that $[X, Y]$ is countable whenever X and Y are triangulable. This implies, for example, since $\pi_1(X, x_0) \subset [S^1, X]$, the fundamental group of a triangulable space is countable.

Exercises

1. Suppose that K is an abstract simplicial complex of dimension n. To find a geometric realization of K, we want to identify vertices of K with points in some $\mathbb{R}^N$ in such a way that, whenever $\{v_0, \ldots, v_q\}$ is a simplex in K, then the associated points $\{\mathbf{v}_0, \ldots, \mathbf{v}_q\}$ are in general position in $\mathbb{R}^N$. In $\mathbb{R}^{2n+1}$ consider the curve

$$\mathcal{C} = \{(r, r^2, \ldots, r^{2n+1}) \mid r \in \mathbb{R}\}.$$

Using the Vandermonde determinant, any $2n + 2$ distinct points on $\mathcal{C}$ are in general position ([**36**]). Assign to each vertex in K a distinct point on $\mathcal{C}$. Since $\dim K = n$, a simplex in K determines at most n points on $\mathcal{C}$ and hence a set in general position. We next worry about intersections of these geometric simplices. Suppose $\{\mathbf{v}_0, \ldots, \mathbf{v}_i, \ldots, \mathbf{v}_{i+k}\}$ and $\{\mathbf{v}_i, \ldots, \mathbf{v}_{i+k}, \ldots, \mathbf{v}_m\}$ are simplices with a shared face. Then $m < 2n + 2$ because $\dim K = n$ and so the union of these sets is in general position. Show that this guarantees that the intersection between these simplices is along a common face alone. Thus we can take an abstract simplicial complex as a geometric simplicial complex without hesitation.

2. Draw a picture (or better yet, make a model) of the first and second barycentric subdivisions of $\mathbf{\Delta}^3$.

3. If K and L are simplicial complexes, their **join**, $K * L$ is the set consisting of the simplices of K, the simplices of L, and the set of 1-simplices $\{\{a, b\} \mid a$ a vertex in K, b a vertex in $L\}$. Show that $K * L$ is a simplicial complex. When $L = \{v_0\}$ and $v_0 \notin K$, show that $K * \{v_0\}$ has CK, the cone on K, as realization.

4. Suppose that $\phi \colon K \to L$ is a simplicial mapping. Suppose that $\psi \colon K \to L$ is a simplicial approximation to $|\phi| \colon |K| \to |L|$. Show that $\psi = \phi$. Thus a simplicial mapping is its own simplicial approximation.

5. Suppose that $f \colon |K| \to |L|$ has a simplicial approximation $\phi \colon K \to L$. Show that sd $\phi \colon$ sd $K \to$ sd L is also a simplicial approximation of f.

6. Prove that composites of contiguous simplicial mappings are contiguous.

7. Suppose K has dimension m and $\phi \colon K \to$ bdy $\mathbf{\Delta}^n$ is a simplicial mapping. If $m < n$, show that $|\phi|$ is null homotopic by showing that the image of $|\phi|$ is not all of $|$ bdy $\mathbf{\Delta}^n|$. This implies that $[S^m, S^n]$ has cardinality one for $m < n$.

Chapter 11

Homology

A complex is a particular type of partially ordered set with complementary properties designed to carry an algebraic superstructure, its homology theory. Complexes thus appear as the tool par excellence for the application of algebraic methods to topology.

SOLOMON LEFSCHETZ, 1942

Simplicial complexes enjoy good topological properties. Their combinatorial structure is sufficiently rich via subdivision to capture the continuous mappings between realizations of complexes up to homotopy. In Chapter 10 we developed these connections between the combinatorial and the continuous. In this chapter we develop the combinatorial structure further by defining algebraic structures associated to a complex that will be found to give topological invariants. These invariants lead to a proof of the topological Invariance of Dimension which is a generalization of the argument in Chapter 8 in which the fundamental group played the key role for the case $(2, n)$.

The algebraic structures will be finite-dimensional vector spaces over the field with two elements, $\mathbb{F}_2 \cong \mathbb{Z}/2\mathbb{Z}$. Let's set some notation: If S is any finite set, then $\mathbb{F}_2[S]$ denotes the vector space over $\mathbb{F}_2$ with S as basis, that is, the set of all formal sums $\sum_{s \in S} a_s s$, where

$a_s \in \mathbb{F}_2$. The sum of two such formal sums is given by

$$\sum_{s \in S} a_s s + \sum_{s \in S} b_s s = \sum_{s \in S} (a_s + b_s) s.$$

Multiplication by a scalar $c \in \mathbb{F}_2$ is given by $c \sum_{s \in S} a_s s = \sum_{s \in S} c a_s s$. The reader can check that these operations make $\mathbb{F}_2[S]$ a vector space. If S and T are finite sets, and $f \colon S \to \mathbb{F}_2[T]$ is a function, then f induces a linear mapping $f_* \colon \mathbb{F}_2[S] \to \mathbb{F}_2[T]$, given by

$$f_* \left(\sum_{s \in S} a_s s \right) = \sum_{s \in S} a_s f(s).$$

Since a linear mapping is determined by its values on a basis of the domain, this construction gives every linear mapping between $\mathbb{F}_2[S]$ and $\mathbb{F}_2[T]$.

The quotient construction of a vector space by a linear subspace (Chapter 1) will come up later, and we recall it here. Suppose W is a linear subspace of a vector space V. The **quotient vector space** V/W is the set of equivalence classes of vectors in V under the equivalence relation $v \sim v'$ if $v' - v \in W$. We denote the equivalence class of $v \in V$ by $[v]$ or $v + W$. The addition and multiplication by a scalar on V/W are given by $(v + W) + (v' + W) = (v + v') + W$ and $c(v + W) = cv + W$. When V is finite dimensional, $\dim V/W = \dim V - \dim W$.

In Chapter 9 we associated to a grating $\mathcal{G}$ the vector space of i-chains, $C_i(\mathcal{G}) = \mathbb{F}_2[E_i(\mathcal{G})]$. We can generalize that construction to a simplicial complex: Suppose K is a simplicial complex (geometric or abstract). Partition K into disjoint subsets that contain only nondegenerate simplices of a fixed dimension:

$$K_p = \{ S \in K \mid \dim S = p \text{ and } S \text{ is nondegenerate} \}.$$

The index p varies from zero to the dimension of K. Each K_p is a finite set which forms the basis for the p-**chains** on K,

$$C_p(K; \mathbb{F}_2) = \mathbb{F}_2[K_p], \text{ the vector space over } \mathbb{F}_2 \text{ with basis } K_p.$$

A typical element of $C_p(K; \mathbb{F}_2)$ is a sum $S_1 + S_2 + \cdots + S_l$, where each S_i is a nondegenerate p-simplex in K. When working over $\mathbb{F}_2$, recall that $S + S = 2 \cdot S = 0 \cdot S = \mathbf{0}$ in $C_p(K; \mathbb{F}_2)$.

A simplicial mapping $\phi \colon K \to L$ induces a linear mapping $\phi_* \colon C_p(K; \mathbb{F}_2) \to C_p(L; \mathbb{F}_2)$ defined on a p-simplex $S = \{v_0, \ldots, v_p\}$ by

$$\phi_*(\{v_0, \ldots, v_p\})$$

$$= \begin{cases} \{\phi(v_0), \ldots, \phi(v_p)\} & \text{if } \{\phi(v_0), \ldots, \phi(v_p)\} \text{ is nondegenerate in } L, \\ 0 & \text{if } \{\phi(v_0), \ldots, \phi(v_p)\} \text{ is degenerate in } L, \end{cases}$$

and defined on a chain $c = S_1 + \cdots + S_l$ by

$$\phi_*(c) = \phi_*(S_1 + \cdots + S_l) = \phi_*(S_1) + \cdots + \phi_*(S_l).$$

If we have two simplicial mappings $\phi \colon K \to L$ and $\psi \colon L \to M$, then the composite $\psi \circ \phi \colon K \to M$ induces a mapping $(\psi \circ \phi)_* \colon C_p(K; \mathbb{F}_2) \to C_p(M; \mathbb{F}_2)$ which satisfies the equation $(\psi \circ \phi)_* = \psi_* \circ \phi_*$.

In Chapter 10 we introduced the face of a p-simplex $S = \{v_0, \ldots, v_p\}$ opposite a vertex v_i, given by the subset

$$\partial_i(S) = \{v_0, \ldots, \widehat{v_i}, \ldots, v_p\} \subset S$$

(the vertex under the hat is omitted). Notice that if S is nondegenerate, then so is $\partial_i S$. Define a mapping $\partial \colon K_p \to C_{p-1}(K; \mathbb{F}_2)$ by summing all of the $(p-1)$-faces of a p-simplex. The extension of ∂ to a linear mapping $C_p(K; \mathbb{F}_2) \to C_{p-1}(K; \mathbb{F}_2)$ is called the **boundary homomorphism**:

$$\partial \colon C_p(K; \mathbb{F}_2) \to C_{p-1}(K; \mathbb{F}_2)$$

given by

$$\partial(S) = \sum_{i=0}^{p} \partial_i(S), \text{ for } S \in K_p.$$

Recall from Chapter 10 that bdy $\Delta^n[S] = \bigcup_{i=0}^{p} \Delta^{n-1}[\partial_i(S)]$. The boundary homomorphism ∂ is an algebraic version of bdy, the topological boundary operation.

The main algebraic properties of the boundary homomorphism are the following:

Proposition 11.1. *If $\phi \colon K \to L$ is a simplicial mapping, then*

$$\partial \circ \phi_* = \phi_* \circ \partial \colon C_p(K; \mathbb{F}_2) \to C_{p-1}(L; \mathbb{F}_2).$$

Furthermore, the composite $\partial \circ \partial \colon C_p(K; \mathbb{F}_2) \to C_{p-2}(K; \mathbb{F}_2)$ is the zero mapping.

Proof. It suffices to check these equations for elements in a basis. Suppose that $S = \{v_0, \ldots, v_p\}$ is a nondegenerate p-simplex in K. Then

$$\partial \circ \phi_*(S) = \partial(\{\phi(v_0), \ldots, \phi(v_p)\})$$
$$= \sum_{i=0}^{p} \{\phi(v_0), \ldots, \widehat{\phi(v_i)}, \ldots, \phi(v_p)\}$$
$$= \sum_{i=0}^{p} \phi_*(\{v_0, \ldots, \widehat{v_i}, \ldots, v_p\})$$
$$= \phi_* \left(\sum_{i=0}^{p} \{v_0, \ldots, \widehat{v_i}, \ldots, v_p\} \right)$$
$$= \phi_* \circ \partial(S).$$

Next, we compute $\partial \circ \partial(S)$.

$$\partial(\partial(S)) = \partial \left(\sum_{i=0}^{p} \{v_0, \ldots, \widehat{v_i}, \ldots, v_p\} \right)$$
$$= \sum_{j<i} \sum_{i=0}^{p} \{v_0, \ldots, \widehat{v_j}, \ldots, \widehat{v_i}, \ldots, v_p\}$$
$$+ \sum_{j>i} \sum_{i=0}^{p} \{v_0, \ldots, \widehat{v_i}, \ldots, \widehat{v_j}, \ldots, v_p\}.$$

Notice that the $(p-2)$-simplex $\{v_0, \ldots, \widehat{v_k}, \ldots, \widehat{v_l}, \ldots, v_p\}$ appears twice for each pair $k < l$, once in each sum, and so $\partial(\partial(S)) = \mathbf{0}$. $\square$

The boundary homomorphism determines certain linear subspaces of $C_p(K; \mathbb{F}_2)$: the **space of p-cycles**,

$$Z_p(K) = \ker(\partial \colon C_p(K; \mathbb{F}_2) \to C_{p-1}(K; \mathbb{F}_2))$$
$$= \{c \in C_p(K; \mathbb{F}_2) \mid \partial(c) = \mathbf{0}\},$$

and the **space of p-boundaries**,

$$B_p(K) = \partial(C_{p+1}(K; \mathbb{F}_2)) = \mathrm{im}(\partial \colon C_{p+1}(K; \mathbb{F}_2) \to C_p(K; \mathbb{F}_2))$$
$$= \{b \in C_p(K; \mathbb{F}_2) \mid b = \partial(c), \text{ for some } c \in C_{p+1}(K; \mathbb{F}_2)\}.$$

The relation $\partial \circ \partial = 0$ implies the inclusion $B_p(K) \subset Z_p(K)$.

For a p-simplex S, the boundary $\partial(S)$ is a cycle that is the sum of the faces $\partial_i(S)$ and together these make up the boundary of $\Delta^p[S]$. When faces come together like this, but the simplex whose boundary they form is absent, we get a 'p-dimensional hole' in the realization of the simplicial complex. The vector space of the essential cycles— holes not filled in as the boundary of a higher-dimensional simplex— is algebraically expressed as the quotient vector space $Z_p(K)/B_p(K)$. This is the homology in dimension p of a simplicial complex.

Definition 11.2. The pth **homology** (mod 2) of a simplicial complex K is the quotient vector space for $p > 0$ given by

$$H_p(K; \mathbb{F}_2) = Z_p(K)/B_p(K).$$

When $p = 0$, define $H_0(K; \mathbb{F}_2) = C_0(K; \mathbb{F}_2)/B_0(K)$.

To illustrate the definition, we compute the homology of the one-point complex, $\Delta^0 = \{v\}$. In this case, the 0-chains have a single vertex $\{v\}$ for a basis, and the boundary homomorphism is zero. Since there are no other simplices, $H_0(\Delta^0; \mathbb{F}_2) = \mathbb{F}_2[\{v\}]$, and $H_p(\Delta^0; \mathbb{F}_2) = \{0\}$ for $p > 0$.

A slightly more complicated computation is the homology of a 1-simplex, $\Delta^1 \cong \Delta^1[S]$, where $S = \{e_0, e_1\}$: the chains and boundary homomorphisms may be assembled into a sequence of vector spaces and linear mappings:

$$\{0\} \to C_1(\Delta^1; \mathbb{F}_2) \xrightarrow{\partial} C_0(\Delta^1; \mathbb{F}_2) \to \{0\}$$

$$\Longleftrightarrow \quad \{0\} \to \mathbb{F}_2[\{S\}] \xrightarrow{\partial} \mathbb{F}_2[\{e_0, e_1\}] \to \{0\}.$$

Since $\partial(S) = e_0 + e_1 \neq \mathbf{0}$, there is no kernel in dimension one, and the zero boundaries are given by $B_0(\Delta^1) = \mathbb{F}_2[\{e_0 + e_1\}]$. Thus $H_0(\Delta^1; \mathbb{F}_2) \cong \mathbb{F}_2[\{[e_0]\}]$, where the equivalence class $[e_0] = e_0 + \mathbb{F}_2[\{e_0 + e_1\}]$ is the coset of e_0 in the quotient $\mathbb{F}_2[\{e_0, e_1\}]/\mathbb{F}_2[\{e_0 + e_1\}]$.

To generalize this computation to $H_p(\Delta^n; \mathbb{F}_2)$ for all n and p, we introduce a linear mapping fashioned from the combinatorics of a simplex. Let $S = \{v_0, \dots, v_n\}$ denote a nondegenerate n-simplex. Consider the linear mapping $i_{v_n}: C_p(\Delta^n[S]; \mathbb{F}_2) \to C_{p+1}(\Delta^n[S]; \mathbb{F}_2)$ given on the basis by

$$i_{v_n}(\{v_{i_0}, \dots, v_{i_p}\})$$

$$= \begin{cases} \{v_{i_0}, \dots, v_{i_p}, v_n\} & \text{if } \{v_{i_0}, \dots, v_{i_p}, v_n\} \text{ is nondegenerate,} \\ \mathbf{0} & \text{otherwise.} \end{cases}$$

If $\{v_{i_0}, \ldots, v_{i_p}\}$ is a nondegenerate p-simplex in $\Delta^n[S]$, $p > 0$, and $v_n \neq v_{i_k}$ for all k, we can compute

$$(\partial \circ i_{v_n} + i_{v_n} \circ \partial)(\{v_{i_0}, \ldots, v_{i_p}\})$$

$$= \partial(\{v_{i_0}, \ldots, v_{i_p}, v_n\}) + i_{v_n}\left(\sum_{r=0}^{p}\{v_{i_0}, \ldots, \widehat{v_{i_r}}, \ldots, v_{i_p}\}\right)$$

$$= \{v_{i_0}, \ldots, v_{i_p}\} + \sum_{r=0}^{p}\{v_{i_0}, \ldots, \widehat{v_{i_r}}, \ldots, v_{i_p}, v_n\}$$

$$+ \sum_{r=0}^{p}\{v_{i_0}, \ldots, \widehat{v_{i_r}}, \ldots, v_{i_p}, v_n\}$$

$$= \{v_{i_0}, \ldots, v_{i_p}\}.$$

When $S = \{v_{i_0}, \ldots, v_{i_{p-1}}, v_n\}$, then $(\partial \circ i_{v_n} + i_{v_n} \circ \partial)(S) = S + U$, where U is a sum of degenerate $(p+1)$-simplices which we take to be $\mathbf{0} \in C_{p+1}(K; \mathbb{F}_2)$. It follows that $\partial \circ i_{v_n} + i_{v_n} \circ \partial = \mathrm{id}$, and if z is a p-cycle, then

$$z = (\partial \circ i_{v_n} + i_{v_n} \circ \partial)(z) = \partial(i_{v_n}(z)) \in B_p(K).$$

Hence, for $p > 0$, $Z_p(K) \subset B_p(K) \subset Z_p(K)$ and so $H_p(\Delta^n[S]; \mathbb{F}_2) = \{\mathbf{0}\}$.

To compute $H_0(\Delta^n[S]; \mathbb{F}_2)$, notice that $\partial(i_{v_n}(v)) = v + v_n$ while $i_{v_n}(\partial(v)) = \mathbf{0}$. The equation $\partial \circ i_{v_n} + i_{v_n} \circ \partial = \mathrm{id}$ does not hold, but we can deduce that $v_n + B_0(\Delta^n[S]) = v_i + B_0(\Delta^n[S])$ for all i. Since $Z_0(\Delta^n[S]) = C_0(\Delta^n[S]; \mathbb{F}_2) = \mathbb{F}_2[\{v_0, \ldots, v_q\}]$, we have

$$H_0(\Delta^n[S]; \mathbb{F}_2) \cong C_0(\Delta^n[S]; \mathbb{F}_2)/\mathbb{F}_2[\{v + v' \mid v \neq v', v, v' \in S\}]$$

$$\cong \mathbb{F}_2[\{v_n + B_0(\Delta^n[S])\}].$$

Notice that the homology of an n-simplex is isomorphic to the homology of a 0-simplex for all n.

We collect the vector spaces of p-chains on Δ^n for all p, together with the boundary homomorphisms, to get a sequence of linear mappings

$$\{\mathbf{0}\} \to C_n(\Delta^n; \mathbb{F}_2) \xrightarrow{\partial} C_{n-1}(\Delta^n; \mathbb{F}_2)$$

$$\xrightarrow{\partial} \cdots \xrightarrow{\partial} C_1(\Delta^n; \mathbb{F}_2) \xrightarrow{\partial} C_0(\Delta^n; \mathbb{F}_2) \to \{\mathbf{0}\}.$$

From the formula $\partial \circ i_{v_n} + i_{v_n} \circ \partial = \mathrm{id}$, we found that, for $p > 0$, $Z_p(\Delta^n) = B_p(\Delta^n)$. In general, we say that a sequence of linear mappings $V \xrightarrow{a} W \xrightarrow{b} U$ is **exact at** W if $\ker b = \mathrm{im}\, a$. In the case of the sequence of chains on $\mathbf{\Delta}^n$, it is exact at $C_i(\mathbf{\Delta}^n; \mathbb{F}_2)$ for $1 \leq i \leq n$. In fact, the pth homology of a simplicial complex, $H_p(K; \mathbb{F}_2) = Z_p(K)/B_p(K)$, measures the failure of the sequence of boundary homomorphisms to be exact at $C_p(K; \mathbb{F}_2)$.

The exactness of the sequence of chains on $\mathbf{\Delta}^n$ gives a method for the computation of $H_p(\text{bdy } \mathbf{\Delta}^n; \mathbb{F}_2)$. The set of simplices of bdy $\mathbf{\Delta}^n$ contains all of the simplices of $\mathbf{\Delta}^n$ $\textit{except}$ the n-simplex $\{\mathbf{e}_0, \ldots, \mathbf{e}_n\}$. We can present the sequence of vector spaces of chains and boundary homomorphisms for bdy $\mathbf{\Delta}^n$ as

$$\{\mathbf{0}\} \to C_{n-1}(\mathbf{\Delta}^n; \mathbb{F}_2) \xrightarrow{\partial} C_{n-2}(\mathbf{\Delta}^n; \mathbb{F}_2)$$

$$\xrightarrow{\partial} \cdots \xrightarrow{\partial} C_1(\mathbf{\Delta}^n; \mathbb{F}_2) \xrightarrow{\partial} C_0(\mathbf{\Delta}^n; \mathbb{F}_2) \to \{\mathbf{0}\}.$$

We know that the sequence is exact at $C_i(\mathbf{\Delta}^n; \mathbb{F}_2)$ for $1 \leq i \leq n - 2$, that the sequence used to be exact at $C_{n-1}(\mathbf{\Delta}^n; \mathbb{F}_2)$, and that $C_n(\mathbf{\Delta}^n; \mathbb{F}_2) = \mathbb{F}_2[\{\mathbf{e}_0, \ldots, \mathbf{e}_n\}]$. In the sequence for bdy $\mathbf{\Delta}^n$, the vector space of $(n-1)$-cycles $Z_{n-1}(\text{bdy } \mathbf{\Delta}^n)$ has dimension one. Since $B_{n-1}(\text{bdy } \mathbf{\Delta}^n) = \{\mathbf{0}\}$, we deduce that

$$H_p(\text{bdy } \mathbf{\Delta}^n; \mathbb{F}_2) \cong \begin{cases} \mathbb{F}_2, & \text{if } p = 0 \text{ or } p = n - 1, \\ \{\mathbf{0}\}, & \text{otherwise.} \end{cases}$$

As we showed in Chapter 10, the realization $|\text{bdy } \mathbf{\Delta}^n|$ is homeomorphic to S^{n-1}. Later we will show how the homology of bdy $\mathbf{\Delta}^n$ can be associated to the topological space S^{n-1}.

To a simplicial complex K we can associate a number based on the combinatorial data of the simplices: Recall the subsets $K_p \subset K$ given by the nondegenerate p-simplices of K. Since K is a finite set, K_p is finite. Let $n_p = \#K_p$, the cardinality of K_p. The **Euler-Poincaré characteristic** of K is the alternating sum

$$\chi(K) = \sum\nolimits_{p=0}^{d} (-1)^p n_p,$$

where d denotes the dimension of K. This number was introduced by Euler in 1750 in a letter to CHRISTIAN GOLDBACH (1690–1764). Euler's formula, $v - e + f = 2$, applies to two-dimensional polyhedra that are homeomorphic to the sphere, but we are getting a little ahead of the story. Here $v = \#$ vertices $= n_0$, $e = \#$ edges $= n_1$, and

$f = \# \text{faces} = n_2$. For example, for the tetrahedron bdy Δ^3 we have $v = 4$, $e = 6$, and $f = 4$.

An extraordinary property of $\chi(K)$ is that it is calculable from the homology.

Theorem 11.3. *If K is a simplicial complex with*

$$\chi(K) = \sum_{p=0}^{d} (-1)^p n_p,$$

then $\chi(K) = \sum_{p=0}^{d} (-1)^p h_p$, where $h_p = \dim_{\mathbb{F}_2} H_p(K; \mathbb{F}_2)$.

Proof. By definition $n_p = \#K_p = \dim_{\mathbb{F}_2} C_p(K; \mathbb{F}_2)$. There are other numbers associated to the chains via the boundary operator. Let

$$z_p = \dim_{\mathbb{F}_2} \ker(\partial\colon C_p(K; \mathbb{F}_2) \to C_{p-1}(K; \mathbb{F}_2)),$$
$$b_p = \dim_{\mathbb{F}_2} \operatorname{im}(\partial\colon C_{p+1}(K; \mathbb{F}_2) \to C_p(K; \mathbb{F}_2)).$$

By definition $h_p = \dim_{\mathbb{F}_2} H_p(K; \mathbb{F}_2) = \dim_{\mathbb{F}_2} Z_p(K)/B_p(K) = z_p - b_p$. The fundamental identity from linear algebra for linear mappings, that the dimension of the domain of a mapping is equal to the dimension of its kernel plus the dimension of its image, implies that $n_p = z_p + b_{p-1}$. Manipulating these identities, we have

$$\chi(K) = \sum_{p=0}^{d} (-1)^p n_p = \sum_{p=0}^{d} (-1)^p (z_p + b_{p-1})$$
$$= (-1)^d (z_d + b_{d-1}) + (-1)^{d-1}(z_{d-1} + b_{d-2})$$
$$+ \cdots + (-1)(z_1 + b_0) + z_0$$
$$= (-1)^d z_d + (-1)^{d-1}(z_{d-1} - b_{d-1})$$
$$+ \cdots + (-1)(z_1 - b_1) + (z_0 - b_0)$$
$$= (-1)^d h_d + (-1)^{d-1} h_{d-1} + \cdots + (-1)h_1 + h_0$$
$$= \sum_{p=0}^{d} (-1)^p h_p.$$

Thus, the number $\chi(K)$ is calculable from the homology of K. $\square$

Poincaré generalized Euler's formula by this argument in an 1895 paper [**66**] that established the importance of this circle of ideas.

Homology and simplicial mappings

Suppose $\phi\colon K \to L$ is a simplicial mapping. Then ϕ induces a linear mapping of chains, $\phi_*\colon C_p(K;\mathbb{F}_2) \to C_p(L;\mathbb{F}_2)$, for which $\partial \circ \phi_* = \phi_* \circ \partial$. Suppose $[c] = c + B_p(K)$ denotes an element in $H_p(K;\mathbb{F}_2)$. Then $c \in Z_p(K)$, that is, $\partial(c) = \mathbf{0}$, and $\partial(\phi_*(c)) = \phi_*(\partial(c)) = \mathbf{0}$, so $\phi_*(c)$ is an element of $Z_p(L)$. If $c - c' \in B_p(K)$, then $\phi_*(c - c') = \phi_*(\partial(u)) = \partial(\phi_*(u))$, for some $u \in C_{p+1}(K;\mathbb{F}_2)$, and so $\phi_*(c) + B_p(L) = \phi_*(c') + B_p(L)$. Thus we can define

$$H(\phi)\colon H_p(K;\mathbb{F}_2) \to H_p(L;\mathbb{F}_2) \text{ by } H(\phi)(c+B_p(K)) = \phi_*(c) + B_p(L).$$

It follows from the properties of the induced mappings on chains that if $\psi\colon L \to M$ is another simplicial mapping, then $H(\psi \circ \phi) = H(\psi) \circ H(\phi)$. We note also that the identity mapping id$\colon K \to K$ induces the identity mapping $H(\text{id}) = \text{id}\colon H_p(K;\mathbb{F}_2) \to H_p(K;\mathbb{F}_2)$ for all p.

Although there are only finitely many simplicial mappings $\phi\colon K \to L$, there can be other linear mappings $C_p(K;\mathbb{F}_2) \to C_q(L;\mathbb{F}_2)$, which, like i_{v_n}, are defined using the features of simplices which make up the bases. The following notion was introduced by Lefschetz [48].

Definition 11.4. Given two simplicial mappings ϕ and $\psi\colon K \to L$, there is a **chain homotopy** between them if there is a linear mapping $h\colon C_p(K;\mathbb{F}_2) \to C_{p+1}(L;\mathbb{F}_2)$ for each p which satisfies

$$\partial \circ h + h \circ \partial = \phi_* + \psi_*.$$

Theorem 11.5. *If there is a chain homotopy between ϕ and ψ, then $H(\phi) = H(\psi)$.*

Proof. Suppose $[c] = c + B_p(K) \in H_p(K;\mathbb{F}_2)$. Then

$$\partial \circ h(c) + h \circ \partial(c) = \phi_*(c) + \psi_*(c).$$

Since $\partial(c) = \mathbf{0}$, $\phi_*(c) + \psi_*(c) = \partial(h(c)) \in B_p(L)$, that is, $\phi_*(c) + B_p(L) = \psi_*(c) + B_p(L)$ and $H(\phi)([c]) = H(\psi)([c])$. $\square$

An important source of chain homotopies is the combinatorial notion of contiguous simplicial mappings. Recall that simplicial mappings $\phi, \psi\colon K \to L$ are *contiguous* if, for any simplex $S \in K$, we have $\phi(S) \cup \psi(S)$ is a simplex in L.

Corollary 11.6. *If ϕ and $\psi \colon K \to L$ are simplicial mappings, and ϕ is contiguous to ψ, then $H(\phi) = H(\psi) \colon H_p(K; \mathbb{F}_2) \to H_p(L; \mathbb{F}_2)$ for all p.*

Proof. Define the linear mapping $h \colon C_p(K; \mathbb{F}_2) \to C_{p+1}(L; \mathbb{F}_2)$ determined on the basis by

$$h(\{v_0, \ldots, v_p\}) = \sum_{i=0}^{p} \{\phi(v_0), \ldots, \phi(v_i), \psi(v_i), \ldots, \psi(v_p)\},$$

where we substitute the zero element whenever we have a degenerate simplex in the sum. Since ϕ and ψ are contiguous, each summand of $h(\{v_0, \ldots, v_p\})$ is a simplex in L.

Then we can compute:

$$(\partial \circ h)(T) = \partial(h(T)) = \partial\left(\sum_{i=0}^{p} \{\phi(v_0), \ldots, \phi(v_i), \psi(v_i), \ldots, \psi(v_p)\}\right)$$

$$= \sum_{i=0}^{p}\sum_{j \leq i} \{\phi(v_0), \ldots, \widehat{\phi(v_j)}, \ldots, \phi(v_i), \psi(v_i), \ldots, \psi(v_p)\}$$

$$+ \sum_{i=0}^{p}\sum_{j \geq i} \{\phi(v_0), \ldots, \phi(v_i), \psi(v_i), \ldots, \widehat{\psi(v_j)}, \ldots, \psi(v_p)\},$$

$$(h \circ \partial)(T) = h(\partial(T)) = h\left(\sum_{i=0}^{p} \{v_0, \ldots, \widehat{v_i}, \ldots, v_p\}\right)$$

$$= \sum_{i=0}^{p}\sum_{j < i} \{\phi(v_0), \ldots, \phi(v_j), \psi(v_j), \ldots, \widehat{\psi(v_i)}, \ldots, \psi(v_p)\}$$

$$+ \sum_{i=0}^{p}\sum_{j > i} \{\phi(v_0), \ldots, \widehat{\phi(v_i)}, \ldots, \phi(v_j), \psi(v_j), \ldots, \psi(v_p)\}.$$

The differences between these expressions are the inequalities $j < i$ and $j \leq i$, and $j > i$ and $j \geq i$. In the sum for $\partial(h(T))$ the summands that do not appear in $h(\partial(T))$ are given by the condition $i = j$:

$$\sum_{i=0}^{p} \{\phi(v_0), \ldots, \phi(v_{i-1}), \psi(v_i), \ldots, \psi(v_p)\}$$

$$+ \{\phi(v_0), \ldots, \phi(v_i), \psi(v_{i+1}), \ldots, \psi(v_p)\}.$$

Each entry appears twice in this sum, except when $i = 0$ and $i = p$, leaving

$$\{\phi(v_0), \ldots, \phi(v_p)\} + \{\psi(v_0), \ldots, \psi(v_p)\} = (\phi_* + \psi_*)(\{v_0, \ldots, v_p\}).$$

All of the summands in $h(\partial(T))$ are cancelled by the rest of the summands of $\partial(h(T))$ and so we have $\partial \circ h + h \circ \partial = \phi_* + \psi_*$, a chain homotopy between ϕ and ψ. By Theorem 11.5, $H(\phi) = H(\psi)$. $\square$

By Lemma 10.19, Corollary 11.6 implies the following:

Corollary 11.7. *If ϕ and $\psi\colon K \to L$ are simplicial approximations of a continuous mapping $f\colon |K| \to |L|$, then $H(\phi) = H(\psi)\colon H_p(K;\mathbb{F}_2) \to H_p(L;\mathbb{F}_2)$, for all p.*

Since a single continuous mapping might have numerous simplicial approximations, when the domain and codomain are held fixed, the induced mappings on homology by these approximations are the same.

Topological invariance

So far we have associated a sequence of vector spaces over $\mathbb{F}_2$ to a simplicial complex. To fashion a tool for the investigation of topological questions, we want to associate homology vector spaces and linear mappings to spaces and continuous mappings. It would be nice to do this for general topological spaces, but it is not clear that it is possible to associate a finite simplicial complex to each space (it isn't [**73**]). We restrict our attention to triangulable spaces, that is, spaces X for which there is a simplicial complex K with X homeomorphic to $|K|$. For such spaces it would be natural to define $H_p(X;\mathbb{F}_2) = H_p(K;\mathbb{F}_2)$. However, a triangulable space can be homeomorphic to many different simplicial complexes. For example, the sphere S^2 is homeomorphic to the tetrahedron, the octohedron, and the icosahedron. It is also the case (Theorem 10.12) that we can subdivide a simplicial complex without changing its realization. How does homology behave under subdivision?

We also want to associate to a continuous mapping $f\colon X \to Y$ a linear mapping $H(f)\colon H_p(X;\mathbb{F}_2) \to H_p(Y;\mathbb{F}_2)$ for each $p \geq 0$. The natural guess is to take a simplicial approximation $\phi\colon \mathrm{sd}^N K \to L$ and define $H(f) = H(\phi)$. This definition is nearly well defined because two simplicial approximations to the same mapping are contiguous. However, simplicial approximations to a single mapping can be constructed for which a different number of barycentric subdivisions might be needed, or a different choice of representing simplicial complexes might have been made and so it is not immediate that we have a good definition.

To alleviate some of the problems here, we loosen some of the foundations to allow a new precision. To allow different choices of a simplicial complex with realization homeomorphic to X we can define $H_p(X; \mathbb{F}_2)$ *up to isomorphism*, that is, do not associate a particular vector space to X and p, but an equivalence class of vector spaces in which a choice of simplicial complex determines a representative. The equivalence relation is the notion of isomorphism, that is, we say that vector spaces V and V' are equivalent if there is a linear isomorphism $\alpha \colon V \to V'$ between them. This relation on any set of vector spaces is reflexive, symmetric, and transitive. We also define a relation between linear mappings between equivalent vector spaces: if $\phi \colon V \to W$ and $\phi' \colon V' \to W'$ are linear mappings, and V is isomorphic to V' and W is isomorphic to W', then we say that ϕ is equivalent to ϕ' if there is a diagram of linear mappings

$$
\begin{array}{ccc}
V & \xrightarrow{\ \phi\ } & W \\
\downarrow{\scriptstyle \alpha} & & \downarrow{\scriptstyle \alpha'} \\
V' & \xrightarrow[\ \phi'\ ]{} & W'
\end{array}
$$

that is *commutative*, that is, $\alpha' \circ \phi = \phi' \circ \alpha$ and α and α' are isomorphisms. Once again, this relation is reflexive, symmetric, and transitive and so we can take linear mappings defined up to isomorphism as equivalence classes under this relation. Although we have loosened up how we associate vector spaces and linear mappings to spaces and continuous mappings, certain linear algebraic invariants remain meaningful, such as the dimension of equivalent vector spaces, and the rank of equivalent linear mappings.

With this notion of equivalence in mind, we establish the well-definedness of the proposed definitions. The central problem that needs resolution is the comparison of the homology of two simplicial complexes with the homeomorphic realizations. As a start, let's consider the relation between the homology of a space and its barycentric subdivision; by Theorem 10.12 we know that $|\operatorname{sd} K| = |K|$.

Theorem 11.8. *There is an isomorphism of vector spaces for all* $p \geq 0$

$$
H_p(\operatorname{sd} K; \mathbb{F}_2) \cong H_p(K; \mathbb{F}_2).
$$

Proof. Recall the simplicial mapping $\lambda\colon \text{sd } K \to K$, defined on vertices by "the last vertex,"

$$\lambda(\beta(S)) = \lambda(\beta(\{v_0, \ldots, v_q\})) = v_q.$$

This mapping is a simplicial approximation to the identity $\text{id}\colon |\text{sd } K| \to |K|$. The simplicial mapping λ induces a linear mapping of chains $\lambda_*\colon C_*(\text{sd } K; \mathbb{F}_2) \to C_*(K; \mathbb{F}_2)$.

To construct an inverse mapping to λ_*, we will not define another simplicial mapping, but work explicitly with the chains. Since we have explicit bases for the vector spaces of p-chains, it is possible to define linear mappings that do not necessarily come from a simplicial mapping. One such combinatorial mapping is defined for a fixed choice of vertex $b \in \text{sd } K$, and generalizes the mapping i_{v_n} that figures in the computation of $H_p(\Delta^n[S]; \mathbb{F}_2)$.

Let $i_b\colon C_q(\text{sd } K; \mathbb{F}_2) \to C_{q+1}(\text{sd } K; \mathbb{F}_2)$ be given on the basis by

$$i_b(\{b_0, \ldots, b_q\})$$
$$= \begin{cases} \{b_0, \ldots, b_q, b\}, & \text{when } \{b_0, \ldots, b_q, b\} \text{ is nondegenerate in sd } K, \\ 0, & \text{if } \{b_0, \ldots, b_q, b\} \text{ is degenerate or not in sd } K. \end{cases}$$

The linear mapping i_b has the following properties:

$$\partial(i_b(S)) = S + i_b(\partial(S)) \quad \text{and} \quad \lambda_* \circ i_{\beta(S)} = i_{b_q} \circ \lambda_*, \quad \text{when } S = \{b_0, \ldots, b_q\}.$$

To prove these identities, we compute (where $\lambda(\beta(S_i)) = b_{\omega_i}$)

$$\partial(i_b(S)) = \partial(\{b_0, \ldots, b_q, b\})$$
$$= \{b_0, \ldots, b_q\} + \sum_{i=0}^{q} \{b_0, \ldots, \widehat{b_i}, \ldots, b_q, b\}$$
$$= S + i_b\left(\sum_{i=0}^{q} \{b_0, \ldots, \widehat{b_i} \ldots, b_q\}\right) = S + i_b(\partial(S)),$$

$$\lambda_* \circ i_{\beta(S)}(\{\beta(S_0), \ldots, \beta(S_{q-1})\})$$
$$= \lambda_*(\{\beta(S_0), \ldots, \beta(S_{q-1}), \beta(S)\})$$
$$= \{\lambda(\beta(S_0)), \ldots, \lambda(\beta(S_{q-1})), \lambda(\beta(S))\}$$
$$= \{b_{\omega(0)}, \ldots, b_{\omega(q-1)}, b_q\}$$
$$= i_{b_q}(\{b_{\omega(0)}, \ldots, b_{\omega(q-1)}\})$$
$$= i_{b_q} \circ \lambda_*(\{\beta(S_0), \ldots, \beta(S_{q-1})\}).$$

Using these identities, we define the mapping $\beta_*\colon C_*(K;\mathbb{F}_2) \to C_*(\mathrm{sd}\ K;\mathbb{F}_2)$ by taking a simplex $S \in K$ to the sum of all the simplices in the barycentric subdivision of K that lie in $\Delta^q[S]$. Explicitly we can write

$$\beta_*(S) = \sum\nolimits_{S_0 \prec S_1 \prec \cdots \prec S_{q-1} \prec S} \{\beta(S_0), \beta(S_1), \ldots, \beta(S_{q-1}), \beta(S)\}.$$

However, this expression can be obtained more compactly by the recursive formula:

$$\beta_*(v) = v,\ \text{if } v \text{ is a vertex in } K,$$
$$\beta_*(S) = i_{\beta(S)} \circ \beta_*(\partial(S))\ \text{if } \dim S > 0.$$

For example, $\beta_*(\{a,b\}) = i_{\beta(\{a,b\})}(\beta_*(a+b)) = \{a, \beta(\{a,b\})\} + \{b, \beta(\{a,b\})\}$, that is, the line segment ab is sent to the sum $am+bm$, where m is the midpoint of ab, the barycenter. We leave to the reader the induction argument that identifies the two descriptions of β_*.

In order that β_* defines a mapping on homology, we check the condition that $\partial \circ \beta_* = \beta_* \circ \partial$. On a 1-simplex, $\{a,b\}$, we have that

$$\partial(\beta_*(\{a,b\})) = \partial(\{a, \beta(\{a,b\})\} + \{b, \beta(\{a,b\})\})$$
$$= a + b = \beta_*(a+b) = \beta_*(\partial(\{a,b\})).$$

By induction on the dimension of a simplex, we have

$$\partial(\beta_*(S)) = \partial(i_{\beta(S)}(\beta_*(\partial(S)))) = \beta_*(\partial(S)) + i_{\beta(S)}(\partial\beta_*(\partial(S)))$$
$$= \beta_*(\partial(S)) + i_{\beta(S)}(\beta_*(\partial\partial(S))) = \beta_*(\partial(S)).$$

Any linear mapping $m_*\colon C_p(K;\mathbb{F}_2) \to C_p(L;\mathbb{F}_2)$, defined for all p, that also satisfies $\partial \circ m_* = m_* \circ \partial$, is called a **chain mapping**; furthermore, a chain mapping m_* induces a linear mapping $m_*\colon H_p(K;\mathbb{F}_2) \to H_p(L;\mathbb{F}_2)$ for all p given by $m_*([v]) = [m_*(v)]$. We have showed that β_* is a chain mapping and so it induces a linear mapping for all p, $\beta_*\colon H_p(K;\mathbb{F}_2) \to H_p(\mathrm{sd}\ K;\mathbb{F}_2)$.

To finish the proof of the theorem, we show that β_* and $H(\lambda)$ are inverses. In one direction, we show that $\lambda_* \circ \beta_* = \mathrm{id}$ on $C_p(K;\mathbb{F}_2)$. On vertices $v \in K$, $\lambda_*(\beta_*(v)) = v$. By induction on dimension, we check on a p-simplex $S = \{v_0, \ldots, v_p\}$,

$$\lambda_*(\beta_*(S)) = \lambda_*(i_{\beta(S)}(\beta_*(\partial(S)))$$
$$= i_{v_p}(\lambda_*(\beta_*(\partial(S)))) = i_{v_p}(\partial(S)) = S.$$

The last equation holds because $i_{v_p}(\partial(S)) = S + \partial(i_{v_p}(S))$, and $v_p \in S$ implies that $i_{v_p}(S) = \mathbf{0}$.

We next construct a chain homotopy

$$h \colon C_p(\text{sd } K; \mathbb{F}_2) \to C_{p+1}(\text{sd } K; \mathbb{F}_2)$$

that satisfies

$$\partial \circ h + h \circ \partial = \beta_* \circ \lambda_* + \text{id}.$$

This implies that $\beta_* \circ H(\lambda) = \text{id}$ on $H_p(\text{sd } K; \mathbb{F}_2)$ and establishes that β_* is the inverse of $H(\lambda)$. For $p = 0$, define $h(\beta(S)) = \{v_p, \beta(S)\}$, where $S = \{v_0, \ldots, v_p\}$. Since $\beta_*(\lambda_*(\beta(S))) = \beta_*(v_p) = v_p$, we have

$$\partial(h(\beta(S))) + h(\partial(\beta(S))) = \partial(\{v_p, \beta(S)\}) = v_p + \beta(S)$$
$$= \beta_*(\lambda_*(\beta(S))) + \text{id}(\beta(S)).$$

Note also that

$$h(\beta(S)) = \{v_p, \beta(S)\} \in C_1(\text{sd } \Delta^p[S]; \mathbb{F}_2) \subset C_1(\text{sd } K; \mathbb{F}_2).$$

Suppose, by induction, that we have defined $h \colon C_k(\text{sd } K; \mathbb{F}_2) \to C_{k+1}(\text{sd } K; \mathbb{F}_2)$ for $k < p$. If $\{\beta(S_0), \ldots, \beta(S_k)\} \in C_k(\text{sd } K; \mathbb{F}_2)$, then let $d_k = \dim(S_k)$. By induction, also assume that

$$h(\{\beta(S_0), \ldots, \beta(S_k)\}) \in C_{k+1}(\text{sd } \Delta^{d_k}[S_k]; \mathbb{F}_2) \subset C_{k+1}(\text{sd } K; \mathbb{F}_2),$$

that is, the chains making up the value of h on a simplex in sd K lie in the subdivision of a particular simplex in K. Suppose T is a p-simplex, $T = \{\beta(S_0), \ldots, \beta(S_p)\}$, and $\dim(S_i) = d_i$. Consider the chain in $C_p(\text{sd } K; \mathbb{F}_2)$ given by $\beta_*(\lambda_*(T)) + T + h(\partial(T))$. By induction, we can assume that $h(\partial(T)) \in C_p(\text{sd } \Delta^{d_p}[S_p]; \mathbb{F}_2)$ since the image under h of any $(p-1)$-simplex $\partial_i(T)$ in $\partial(T)$ lies in $C_{p-1}(\text{sd } \Delta^{d_p}[S_p]; \mathbb{F}_2) \oplus C_p(\text{sd } \Delta^{d_p-1}[S_{p-1}]; \mathbb{F}_2) \subset C_p(\text{sd } \Delta^{d_p}[S_p]; \mathbb{F}_2)$. Since $S_0 \prec S_1 \prec \cdots \prec S_p$, we know that $T \in \text{sd } \Delta^{d_p}[S_p]$. Finally, consider

$$\beta_*(\lambda_*(T)) = \beta_*(\{v_{\omega(0)}, \ldots, v_{\omega(p)}\})$$
$$\in C_p(\text{sd } \Delta^p[\{v_{\omega(0)}, \ldots, v_{\omega(p)}\}]; \mathbb{F}_2).$$

Since $v_{\omega(i)}$ lies in $S_i \prec S_p$, we find $\beta_*(\lambda_*(T)) \in C_p(\text{sd } \Delta^{d_p}[S_p]; \mathbb{F}_2)$.

Putting these observations together it follows that the p-chain

$$\beta_*(\lambda_*(T)) + T + h(\partial(T)) \in C_p(\text{sd } \Delta^{d_p}[S_p]; \mathbb{F}_2).$$

The sequence of chains and boundary homomorphisms for sd $\Delta^{d_p}[S_p]$ is exact in dimensions greater than zero because the operator $i_{\beta(S_p)}$: $C_k(\text{sd } \Delta^{d_p}[S_p]; \mathbb{F}_2) \to C_{k+1}(\text{sd } \Delta^{d_p}[S_p]; \mathbb{F}_2)$ satisfies $\partial \circ i_{\beta(S_p)} + \partial \circ i_{\beta(S_p)} = \text{id}$ (the proof is the same as for $\Delta^{d_p}[S_p]$). Furthermore, by induction, we can assume that $\beta_* \circ \lambda_* + \text{id} = h \circ \partial + \partial \circ h$ on $(p-1)$-chains, and so

$$\begin{aligned}
\partial(\beta_* \circ \lambda_* + \text{id} + h \circ \partial) &= \partial \circ \beta_* \circ \lambda_* + \partial + (\partial \circ h) \circ \partial \\
&= \beta_* \circ \lambda_* \circ \partial + \partial + (\beta_* \circ \lambda_* + \text{id} + h \circ \partial) \circ \partial \\
&= \beta_* \circ \lambda_* \circ \partial + \partial + \beta_* \circ \lambda_* \circ \partial + \partial + h \circ \partial \circ \partial = 0.
\end{aligned}$$

Thus

$$\beta_*(\lambda_*(T)) + T + h(\partial(T)) \in Z_p(\text{sd } \Delta^{d_p}[S_p]) = B_p(\text{sd } \Delta^{d_p}[S_p]).$$

Therefore, there is a $(p+1)$-chain $c_T \in C_{p+1}(\text{sd } \Delta^{d_p}[S_p]; \mathbb{F}_2) \subset C_{p+1}(\text{sd } K; \mathbb{F}_2)$ with $\partial(c_T) = \beta_*(\lambda_*(T)) + T + h(\partial(T))$. Define $h(T) = c_T$. Carry out this construction for each $T \in K_p$ and extend linearly to define $h \colon C_p(\text{sd } K; \mathbb{F}_2) \to C_{p+1}(\text{sd } K; \mathbb{F}_2)$, satisfying $\beta_* \circ \lambda_* + \text{id} = \partial \circ h + h \circ \partial$, and $h(T) \in C_{p+1}(\text{sd } \Delta^{d_p}[S_p]; \mathbb{F}_2)$.

It now follows from Theorem 11.5 that $\beta_* \circ \lambda_*$ induces the identity on $H_p(\text{sd } K; \mathbb{F}_2)$ and we have proved that $H_p(K; \mathbb{F}_2) \cong H_p(\text{sd } K; \mathbb{F}_2)$, for all p. $\qquad\square$

The trick of restricting and applying the exactness of the sequence of chains and boundary homomorphisms for a subcomplex of a simplicial complex is known generally as the *method of acyclic models*, introduced generally by S. EILENBERG (1913–1998) and J. ZILBER in [**24**].

Since $|\text{sd } K| = |K|$, Theorem 11.8 shows that subdivision does not change the homology up to isomorphism. The Simplicial Approximation Theorem, together with certain properties of simplicial mappings, will imply that the collection of homology vector spaces $\{H_p(K; \mathbb{F}_2) \mid p \geq 0\}$ are topological invariants.

Topological invariance of homology. *Suppose K and L are simplicial complexes with $|K|$ and $|L|$ homeomorphic. Then, for all p, the vector spaces $H_p(K; \mathbb{F}_2)$ and $H_p(L; \mathbb{F}_2)$ are isomorphic.*

Proof*. Suppose $F: |K| \to |L|$ is a homeomorphism with inverse given by $G: |L| \to |K|$. Let $\phi: \mathrm{sd}^N K \to L$ be a simplicial approximation to F and $\gamma: \mathrm{sd}^M L \to K$ a simplicial approximation to G. Then, we can subdivide the simplicial mapping ϕ further to obtain $\mathrm{sd}^M \phi: \mathrm{sd}^{N+M} K \to \mathrm{sd}^M L$ which is also a simplicial approximation to F (Exercise 5, Chapter 10). The composite

$$\mathrm{sd}^{N+M} K \xrightarrow{\mathrm{sd}^M \phi} \mathrm{sd}^M L \xrightarrow{\gamma} K$$

is a simplicial approximation to the identity mapping $|\mathrm{sd}^{N+M} K| \to |K|$. Another approximation of the identity is given by the following composite:

$$\mathrm{sd}^{N+M} K \xrightarrow{\mathrm{sd}^{N+M-1} \lambda} \mathrm{sd}^{N+M-1} K$$

$$\xrightarrow{\mathrm{sd}^{N+M-2} \lambda} \cdots \xrightarrow{\mathrm{sd}^2 \lambda} \mathrm{sd}^2 K \xrightarrow{\mathrm{sd}\, \lambda} \mathrm{sd}\, K \xrightarrow{\lambda} K.$$

The proof of Theorem 11.8 shows that $H(\lambda)$ is an isomorphism between $H_p(\mathrm{sd}\, K; \mathbb{F}_2)$ and $H_p(K; \mathbb{F}_2)$ for all p. We next show that $H(\mathrm{sd}^j \lambda)$ is an isomorphism for all $j \geq 0$. More generally, consider the diagram of simplicial complexes and simplicial mappings:

$$
\begin{array}{ccc}
\mathrm{sd}\, K & \xrightarrow{\mathrm{sd}\, \eta} & \mathrm{sd}\, L \\
\downarrow{\scriptstyle \lambda} & & \downarrow{\scriptstyle \lambda_K} \\
K & \xrightarrow{\eta} & L
\end{array}
$$

Here we define $\lambda_K: \mathrm{sd}\, L \to L$ as a simplicial approximation to the identity that satisfies $\lambda_K(\{\phi(v_0), \ldots, \phi(v_q)\}) = \phi(v_q)$, that is, we complete the diagram in such a way that $\eta \circ \lambda = \lambda_K \circ \mathrm{sd}\, \eta$. When we apply homology to these mappings, we obtain $H(\eta) \circ H(\lambda) = H(\lambda_K) \circ H(\mathrm{sd}\, \eta)$. Since λ and λ_K are simplicial approximations of the identity mapping, they are contiguous and so $H(\lambda_K)$ and $H(\lambda)$ are isomorphisms. Therefore, $H(\eta)$ and $H(\mathrm{sd}\, \eta)$ are equivalent as

---

* Skip this proof on first reading.

linear mappings of vector spaces. From this we deduce that $H(\mathrm{sd}^j\,\lambda)$ is an isomorphism for all $j \geq 0$.

Thus $\gamma\circ\mathrm{sd}^M\;\phi\colon \mathrm{sd}^{N+M}\,K \to K$ and $\lambda\circ(\mathrm{sd}\,\lambda)\circ\cdots\circ(\mathrm{sd}^{N+M-1}\,\lambda)\colon$ $\mathrm{sd}^{N+M}\,K \to K$ are both simplicial approximations to the identity map $|\,\mathrm{sd}^{N+M}\,K| \to |K|$ and so they are contiguous by Lemma 10.19. Thus $H(\gamma)\circ H(\mathrm{sd}^M\,\phi) = H(\lambda)\circ H(\mathrm{sd}\,\lambda)\circ\cdots\circ H(\mathrm{sd}^{N+M-1}\,\lambda)$, which is an isomorphism. It follows that $H(\mathrm{sd}^M\,\phi)$ is one-one and also that $H(\phi)$ is one-one because it is equivalent to $H(\mathrm{sd}^M\,\phi)$.

By the same argument applied to $G \circ F = \mathrm{id}_{|L|}$, we form the composite

$$\mathrm{sd}^{N+M}\,L \xrightarrow{\mathrm{sd}^N\,\gamma} \mathrm{sd}^N\,K \xrightarrow{\phi} L,$$

which is a simplicial approximation to id: $|\,\mathrm{sd}^{N+M}\,L| \to |L|$ and so $H(\phi)\circ H(\mathrm{sd}^N\,\gamma)$ is an isomorphism and so $H(\phi)$ is onto. Thus we have proved that $H(\phi)\colon H_p(\mathrm{sd}^N\,K;\mathbb{F}_2) \to H_p(L;\mathbb{F}_2)$ is an isomorphism, for all p. By Theorem 11.8 and induction, $H_p(K;\mathbb{F}_2)$ is isomorphic to $H_p(\mathrm{sd}^N\,K;\mathbb{F}_2)$. Thus $H_p(K;\mathbb{F}_2) \cong H_p(L;\mathbb{F}_2)$ for all p. $\Box$

Corollary 11.9. *The Euler-Poincaré characteristic is a topological invariant of a triangulable space.*

Proof. Since $\chi(K)$ is calculable from the homology and homology is a topological invariant, we can write $\chi(K) = \chi(|K|)$ and compute the Euler-Poincaré characteristic from any triangulation of $|K|$. $\Box$

The topological invariance of homology allows us to compute $H_p(S^2;\mathbb{F}_2)$ from $H_p(\mathrm{bdy}\,\mathbf{\Delta}^3;\mathbb{F}_2)$. It follows that $\chi(S^2) = \chi(\mathrm{bdy}\,\mathbf{\Delta}^3) = 2$.

We next prove a result known since the time of Euclid. A **Platonic solid** is a polyhedron with realization S^2 and for which all faces are congruent to a regular polygon, and each vertex has the same number of edges meeting there. Familiar examples are the tetrahedron and cube.

Theorem 11.10. *There are only five Platonic solids.*

Proof. A polyhedron P need not be a simplicial complex, since the faces can be polygons not necessarily triangles (consider a soccer ball). However, if we subdivide each constituent polygon into triangles, we get a simplicial complex. The reader can now prove that the Euler-Poincaré characteristic $\chi(P)$, computed as the alternating sum $n_0 - n_1 + n_2$ where P has n_0 vertices, n_1 edges, and n_2 faces, is the same for the subdivided polyhedron, a simplicial complex. Since P has realization homeomorphic to S^2, we know that $\chi(P) = 2$.

Suppose each face has M edges (a regular M-gon) and, at each vertex, N faces meet. This leads to the relation:

$$M n_2/2 = n_1,$$

that is, each of the n_2 faces contributes M edges, but each edge is shared by two faces. It is also the case that

$$N n_0/2 = n_1.$$

Since N faces meet at each vertex, N edges come into each vertex. But each edge has two vertices. Putting these relations into Euler's formula we get

$$
\begin{aligned}
2 &= n_0 - n_1 + n_2 \\
&= (2n_1/N) - n_1 + (2n_1/M) \\
&= n_1((2/N) + (2/M) - 1).
\end{aligned}
$$

It follows that

$$\frac{n_1}{2} = \frac{MN}{2M + 2N - MN}.$$

If $N = 1$ or $N = 2$, there would be a boundary and so the polyhedron would fail to be a sphere. Since a Platonic solid encloses space, $N > 2$. Also $M \geq 3$ since each face is a polygon. Finally, n_1 must be an integer which is at least M.

These facts force $M < 6$. To see this, suppose $M \geq 6$ and $N > 2$. Then $2 - N < 0$ and we have

$$0 < 2M + 2N - MN = 2N + M(2 - N) \leq 2N + 6(2 - N) = 12 - 4N.$$

This implies that $4N < 12$, or that $N < 3$, which is impossible for N an integer and $N > 2$.

Setting $M = 3$ we get $n_1 = 6N/(6 - N)$, which is an integer when $N = 3$, 4, and 5. The case $N = 3$, $M = 3$ is realized by the tetrahedron, $N = 4$ and $M = 3$ is realized by the octahedron, and for $N = 5$, $M = 3$ by the icosahedron.

For $M = 4$ we have $n_1 = 8N/(8 - 2N) = 4N/(4 - N)$, and so $N = 3$ is the only case of interest which is realized by the cube. Finally, for $M = 5$ we have $n_1 = 10N/(10 - 3N)$ and so $N = 3$ is the only possible case, which gives the dodecahedron. $\square$

Since the homology groups of a triangulable space are defined up to isomorphism, the invariants of vector spaces, like dimension, are topological invariants of the space. In the next result, we compare the dimension of one of the homology groups to a topological invariant introduced in Chapter 5.

Theorem 11.11. *If K is a simplicial complex, then* $\dim_{\mathbb{F}_2} H_0(K; \mathbb{F}_2)$ $= \#\pi_0(|K|) =$ *the number of path components of* $|K|$.

Proof. Consider the set K_0 of vertices of K. Define a relation on K_0 given by $v \sim v'$ if there is a 1-chain $c \in C_1(K; \mathbb{F}_2)$ with $\partial(c) = v + v'$. This relation is reflexive, because $\partial(0) = v + v$; it is symmetric since $v + v' = v' + v$; and it is transitive because $\partial(c) = v + v'$ and $\partial(c') = v' + v''$ implies $\partial(c + c') = v + v' + v' + v'' = v + v''$. Let $[K_0]$ denote the set of equivalence classes under this relation. We show that $\#[K_0] = \dim_{\mathbb{F}_2} H_0(K; \mathbb{F}_2)$ and $\#[K_0] = \#\pi_0(|K|)$.

Consider the linear mapping $\mathbb{F}_2[[K_0]] \to H_0(K; \mathbb{F}_2)$ determined by $[v] \mapsto v + B_0(K)$. Since the equivalence relation is defined by the image of the boundary homomorphism, this mapping is well defined. It is onto since every vertex in K lies in some equivalence class in $[K_0]$. We prove that this mapping is an isomorphism. Suppose that we make a choice of vertex in each equivalence class so that $[K_0] = \{[v_1], \ldots, [v_s]\}$. We show that the set of classes $\{v_i + B_0(K) \mid i = 1, \ldots, s\}$ is linearly independent in $H_0(K; \mathbb{F}_2)$. Suppose $v_{i_1} + \cdots + v_{i_r} + B_0(K) = B_0(K)$, that is, $v_{i_1} + \cdots + v_{i_r} = \partial(c)$ for some $c \in C_1(K; \mathbb{F}_2)$. We can write $c = e_1 + \cdots + e_t$ for edges $e_i \in K_1$. Since $v_{i_1} + \cdots + v_{i_r} = \partial(e_1 + \cdots + e_t)$, there is some edge, say e_1, with $\partial(e_1) = v_{i_1} + w_1$ for some vertex w_1. Since $v_{i_1} \sim w_1$, we know that $w_1 \neq v_{i_j}$ for $j = 2, \ldots, s$. It follows that we can replace v_{i_1} with w_1 and write

$$w_1 + v_{i_2} + \cdots + v_{i_r} = \partial(e_2 + \cdots + e_t).$$

By the same argument, we can choose e_2 with $\partial(e_2) = w_1 + w_2$. Once again, $w_1 \sim w_2$ and $w_2 \neq v_{i_j}$ for $j = 2, \ldots, s$. Therefore, $\partial(e_3 + \cdots + e_t) = w_2 + v_{i_2} + \cdots + v_{i_r}$. Continuing in this manner, we get down to $\partial(e_t) = w_{t-1} + v_{i_2} + \cdots + v_{i_r}$, which is impossible since the vertices v_{i_j} and w_{t-1} are not equivalent under the relation. Thus $\#[K_0] = s = \dim_{\mathbb{F}_2} H_0(K; \mathbb{F}_2)$.

To finish the proof, we show that $\#[K_0] = \#\pi_0(|K|)$. First notice that the open star of a vertex, $O_K(v)$, is path-connected. This follows because there is a path joining the vertex v to every point in $O_K(v)$. Recall that the set of path components $\pi_0(|K|)$ is the set of equivalence classes of points in $|K|$ under the relation that two points are equivalent if there is a path in $|K|$ joining them. Denote the equivalence classes under this relation by $\langle x \rangle$. Suppose $[v_i] \in [K_0]$ is a class of vertices under the relation $v_i \sim w$ if there is a 1-chain c with $\partial(c) = v_i + w$. Let $U_i = \bigcup_{w \in [v_i]} O_K(w)$. We show that U_i is a path component of $|K|$ and that $U_i \cap U_j = \emptyset$ when $i \neq j$. If $w, w' \in [v_i]$, then $w \sim w'$ and there is a 1-chain c with $\partial(c) = w + w'$. The 1-chain determines a path joining w to w' and so $O_K(w) \cup O_K(w')$ is path-connected. Continuing with other vertices it follows that U_i is path-connected. If there is a path joining v_i to a point x in $|K|$, then there is a path joining v_i to some vertex v in K, and the path joining v_i to v can be deformed to pass only along edges of K whose sum gives a 1-chain c with $\partial(c) = v_i + v$, that is, $v \in U_i$ and $U_i = \langle v_i \rangle$. Suppose $x \in U_i \cap U_j$. Then there are vertices w and v with $v \sim v_i$ and $w \sim v_j$ and $x \in O_K(v) \cap O_K(w)$. However, this implies that $x \in \Delta^m[S]$ for some m-simplex S in K for which $v, w \in S$. Thus $e = \{v, w\} \prec S$ is an edge with $\partial(e) = v + w$ and so $v \sim w$, which implies $v_i \sim v_j$, a contradiction. Thus $|K|$ is partitioned into disjoint path components $\langle v_1 \rangle = U_1, \ldots, \langle v_s \rangle = U_s$. $\qquad\square$

We now turn to the central question of the book.

Invariance of Dimension for (m, n)**.** *If* $\mathbb{R}^m$ *is homeomorphic to* $\mathbb{R}^n$, *then* $n = m$.

Proof. We make this a question about simplicial complexes by using the one-point compactification (Definition 6.11). If $\mathbb{R}^n$ is homeomorphic to $\mathbb{R}^m$, then their one-point compactifications are homeomorphic. Since $\mathbb{R}^l \cup \{\infty\}$ is homeomorphic to S^l, it follows that $\mathbb{R}^n \cong \mathbb{R}^m$ implies $S^n \cong S^m$.

By the topological invariance of homology, and the homeomorphism $S^n \cong |\,\mathrm{bdy}\,\Delta^{n+1}|$, we have

$$H_p(S^n; \mathbb{F}_2) \cong H_p(\mathrm{bdy}\,\Delta^{n+1}; \mathbb{F}_2) \cong \begin{cases} \mathbb{F}_2, & p = 0, n, \\ \{\mathbf{0}\}, & \text{else.} \end{cases}$$

If $S^n \cong S^m$, then $H_p(S^n; \mathbb{F}_2) \cong H_p(S^m; \mathbb{F}_2)$ for all p and, by our computation of the homology of spheres, this is only possible if $n = m$. $\qquad\square$

The first proofs of this theorem were due to Brouwer [10] and Lebesgue [47]. Brouwer's proof was based on simplicial approximation and used an index, defined generically as the cardinality of the preimage of a point, to obtain a contradiction to the existence of a homeomorphism between $[0,1]^n = [0,1] \times \cdots \times [0,1]$ (n times) and $[0,1]^m$ when $n \neq m$. Lebesgue's first proof was not rigorous, but introduced a point-set definition of dimension that led to the modern development of the subject of dimension theory. An account of these developments can be found in [40] and [17].

Another famous theorem of Brouwer can be proved using homology, generalizing the argument in Theorem 8.7 in which the fundamental group of S^1 played a key role.

The Brouwer Fixed Point Theorem. *If* $e^n = \{\mathbf{x} \in \mathbb{R}^n \mid \|\mathbf{x}\| \leq 1\}$ *denotes the* n-**disk** *and* $f\colon e^n \to e^n$ *is a continuous mapping, then there is a point* $\mathbf{x}_0 \in e^n$ *with* $f(\mathbf{x}_0) = \mathbf{x}_0$, *that is,* e^n *has the fixed point property.*

Proof. Suppose that $f\colon e^n \to e^n$ is a continuous mapping without fixed points. If $\mathbf{y} \in e^n$, then $\mathbf{y} \neq f(\mathbf{y})$. Join $f(\mathbf{y})$ to $\mathbf{y}$ and continue this ray until it meets $S^{n-1} = \mathrm{bdy}\,e^n$ and denote this point by $g(\mathbf{y})$. We can characterize $g(\mathbf{y})$ by $g(\mathbf{y}) = (1-t)f(\mathbf{y}) + t\mathbf{y}$, where $t > 0$ and $\|g(\mathbf{y})\| = 1$. Because we are in a nicely behaved inner product space, the argument for the case of $n = 2$ (Theorem 8.7) carries over exactly to prove that $g\colon e^n \to S^{n-1}$ is continuous. Furthermore, by the definition of g, $g \circ i \colon S^{n-1} \to S^{n-1}$ is the identity when $i \colon S^{n-1} \to e^n$ is the inclusion of the boundary.

Apply homology to this composite $\mathrm{id}_{S^{n-1}} = g \circ i$ to obtain $H(\mathrm{id}_{S^{n-1}})$, an isomorphism, written as $H(g) \circ H(i)$. However, $H_{n-1}(S^{n-1}; \mathbb{F}_2) \neq \{\mathbf{0}\}$ while $H_{n-1}(e^n; \mathbb{F}_2) = \{\mathbf{0}\}$, because e^n is homeomorphic to $\mathbf{\Delta}^n$. Thus, $H(i) \colon H_{n-1}(S^{n-1}; \mathbb{F}_2) \to H_{n-1}(e^n; \mathbb{F}_2)$ is the zero homomorphism $[c] \mapsto \mathbf{0}$. An isomorphism

$$H(\mathrm{id}_{S^{n-1}}) \colon H_{n-1}(S^{n-1}; \mathbb{F}_2) \to H_{n-1}(S^{n-1}; \mathbb{F}_2)$$

cannot be factored as $H(g) \circ ([c] \mapsto \mathbf{0})$, and so a continuous mapping $f \colon e^n \to e^n$ without fixed points cannot exist. $\qquad\square$

The Brouwer Fixed Point Theorem was a significant signpost in the development of topology. The theory of fixed points of mappings plays an important role throughout mathematics and its applications. With more refined notions of homology, deep generalizations of the Brouwer Fixed Point Theorem can be proved. See [**62**] for examples, like the Lefschetz-Hopf Fixed Point Theorem.

In dimension two we proved a case of the Borsuk-Ulam Theorem (Proposition 8.10)—there does not exist a continuous function $f \colon S^2 \to S^1$ with $f(-\mathbf{x}) = -f(\mathbf{x})$ for all $\mathbf{x} \in S^2$. The higher-dimensional version of the Borsuk-Ulam Theorem treats mappings $f \colon S^n \to S^{n-1}$ for which $f(-\mathbf{x}) = -f(\mathbf{x})$. The general setting for this discussion involves the notion of a space with involution.

Definition 11.12. A space X has an **involution** $\nu \colon X \to X$ if ν is continuous and $\nu \circ \nu = \mathrm{id}_X$. If (X, ν) and (Y, μ) are spaces with involution, then an **equivariant mapping** $g \colon X \to Y$ is a continuous mapping satisfying $g \circ \nu = \mu \circ g$.

Consider the antipodal mapping on S^n and on S^{n-1} given by $a(\mathbf{x}) = -\mathbf{x}$. The general Borsuk-Ulam Theorem states that a continuous mapping $f \colon S^n \to S^{n-1}$ cannot be equivariant, that is, $f(a(\mathbf{x})) = a(f(\mathbf{x}))$ does not hold for all $\mathbf{x} \in S^n$.

Assuming this formulation of the Borsuk-Ulam Theorem, we observe an immediate consequence: If we let $F \colon S^n \to \mathbb{R}^n$ be any continuous mapping that satisfies $F(\mathbf{x}) \neq F(-\mathbf{x})$ for all $\mathbf{x} \in S^n$, we can define

$$g(\mathbf{x}) = \frac{F(\mathbf{x}) - F(-\mathbf{x})}{\|F(\mathbf{x}) - F(-\mathbf{x})\|}.$$

Then $g\colon (S^n, a) \to (S^{n-1}, a)$ is an equivariant mapping. By the Borsuk-Ulam Theorem, no such mapping exists, and so there must be a point $\mathbf{x}_0 \in S^n$ with $F(\mathbf{x}_0) = F(-\mathbf{x}_0)$, that is, two antipodal points are mapped to the same point. It follows from this that no subspace of $\mathbb{R}^n$ is homeomorphic to S^n.

We deduce the Borsuk-Ulam Theorem as a corollary of a theorem of Walker [**78**] which deals with the homology of equivariant mappings. Assume that (X, ν) is a space with involution and that X is triangulable. Then there is a simplicial complex K with $|K| \cong X$ and a simplicial mapping $\bar{\nu}\colon K \to K$ with $|\bar{\nu}| \simeq \nu$ and $\bar{\nu} \circ \bar{\nu} = \mathrm{id}_K$. An argument for the existence of K and $\bar{\nu}$ can be made using simplicial approximation. For the sphere, we can do even better. For example, one triangulation of S^2 is the octahedron on which we can write down an explicit simplicial mapping which realizes the antipodal map. Higher dimensional models of this sort exist for every sphere. Note that the antipodal mapping on the sphere has no fixed points. We will assume that a simplicial approximation to the antipodal map can be chosen without fixed points as well, and so any simplex S in L satisfies $\bar{a}(S) \cap S = \emptyset$, where $\bar{a}\colon L \to L$ realizes the antipode on $|L| \cong S^n$.

Theorem 11.13. *If (X, ν) is a triangulable space with involution and $F\colon (X, \nu) \to (S^n, a)$ is an equivariant mapping, then there is a homology class $[c] \in H_j(X; \mathbb{F}_2)$ with $1 \le j \le n$, $[c] \ne \mathbf{0}$, and $H(\nu)([c]) = [c]$. Furthermore, if the least dimension in which this condition holds is $j = n$, then the class $[c]$ can be chosen such that $H(F)([c]) = [u] \ne \mathbf{0}$ in $H_n(S^n; \mathbb{F}_2)$.*

Proof. Let us assume that we have triangulations for (X, ν) and (S^n, a) denoted by $(K, \bar{\nu})$ and $(L, \bar{a})$. Let $\phi\colon K \to L$ be a simplicial equivariant mapping with ϕ a simplicial approximation to F. Let $\theta_K = \mathrm{id}_{K*} + \bar{\nu}_*\colon C_j(K; \mathbb{F}_2) \to C_j(K; \mathbb{F}_2)$ and $\theta_L = \mathrm{id}_{L*} + \bar{a}_*\colon C_j(L; \mathbb{F}_2) \to C_j(L; \mathbb{F}_2)$. Since $\bar{\nu}$ and $\bar{a}$ are simplicial mappings, $\theta_K \circ \partial = \partial \circ \theta_K$

and likewise for θ_L. Also $\theta_K \circ \theta_K = \mathbf{0}$, because

$$(\mathrm{id}_{K*} + \bar{\nu}_*) \circ (\mathrm{id}_{K*} + \bar{\nu}_*) = \mathrm{id}_{K*} + \bar{\nu}_* + \bar{\nu}_* + (\bar{\nu} \circ \bar{\nu})_* = 2\,\mathrm{id}_{K*} + 2\bar{\nu}_* = \mathbf{0},$$

and similarly, $\theta_L \circ \theta_L = \mathbf{0}$.

If there is a class $\mathbf{0} \neq [c] \in H_j(K; \mathbb{F}_2)$ with $H(\bar{\nu})([c]) = [c]$ and $0 < j < n$, then we are done. So, let us assume that if $H(\bar{\nu})([c]) = [c]$, then $[c] = \mathbf{0}$. Notice that $H(\bar{\nu})([c]) = [c]$ if and only if $[\theta_K(c)] = \mathbf{0}$.

Let $h_0 \in L$ denote a vertex. The homology class $[h_0] = h_0 + B_0(L) \in H_0(L; \mathbb{F}_2)$ satisfies $[\theta_L(h_0)] = \mathbf{0}$, since $H_0(L; \mathbb{F}_2)$ has dimension one, and both id_L and $\bar{a}$ induce the identity on $H_0(L; \mathbb{F}_2)$. It follows that there is a 1-chain h_1 with $\partial(h_1) = \theta_L(h_0)$. Notice that

$$\partial(\theta_L(h_1)) = \theta_L(\partial(h_1)) = \theta_L(\theta_L(h_0)) = \mathbf{0}.$$

Since $|L| \cong S^n$, $B_1(L) = Z_1(L)$ and so $\theta_L(h_1) = \partial(h_2)$ for some $h_2 \in C_2(L; \mathbb{F}_2)$. It is also the case that $\theta_L(h_1) \neq \mathbf{0}$. To see this, suppose $h_1 = e_1 + e_2 + \cdots + e_t$. Then we can number the edges e_i with $\partial(e_1) = h_0 + v_1$, $\partial(e_i) = v_{i-1} + v_i$, and $\partial(e_t) = v_{t-1} + \bar{a}_*(h_0)$. If $\theta_L(h_1) = \mathbf{0}$, then we deduce $\bar{a}_*(e_i) = e_{t-i+1}$ from which we find either an edge that is its own antipode, or a pair of edges sharing antipodal vertices. By the assumption that the antipode $\bar{a}$ has no fixed points, we find $\theta_L(h_1) \neq \mathbf{0}$.

We repeat this construction to find $h_j \in C_j(L; \mathbb{F}_2)$, for $1 \leq j \leq n$, with $\partial(h_j) = \theta_L(h_{j-1})$. By the same argument showing $\theta_L(h_1) \neq \mathbf{0}$, we find $\theta_L(h_j) \neq \mathbf{0}$ for $1 \leq j \leq n$. Consider $\theta_L(h_n)$; since $\theta_L(h_n) \neq \mathbf{0}$, $[\theta_L(h_n)]$ generates $H_n(L; \mathbb{F}_2)$. The chains h_j may be thought of as generalized hemispheres.

We have assumed that, if $1 \leq j < n$ and $[c] \in H_j(K; \mathbb{F}_2)$ satisfies $H(\bar{\nu})[c] = [c]$, then $[c] = \mathbf{0}$. We use this to make an analogous construction of classes $c_j \in C_j(K; \mathbb{F}_2)$ with properties like the h_j. Let $c_0 \in K$ be a vertex. Then $[\theta_K(c_0)] = \mathbf{0}$, and so there is a 1-chain c_1 with $\partial(c_1) = \theta_K(c_0)$. The 1-chain $\theta_K(c_1)$ satisfies

$$\partial(\theta_K(c_1)) = \theta_K(\partial(c_1)) = \theta_K(\theta_K(c_0)) = \mathbf{0}.$$

Thus $\theta_K(c_1)$ is a 1-cycle. However, $\theta_K(\theta_K(c_1)) = \mathbf{0}$, so $\theta_K(c_1) = \partial(c_2)$ for some 2-chain c_2. Continuing in this manner, we find chains c_j satisfying $\partial(c_j) = \theta_K(c_{j-1})$ for $1 \leq j \leq n$.

We next define another sequence of chains on L. We know that $h_0 + \phi_*(c_0)$ is a 0-cycle, and so there is a chain u_1 with $\partial(u_1) = h_0 + \phi_*(c_0)$. Consider $h_1 + \phi_*(c_1) + \theta_L(u_1)$. Then

$$
\begin{aligned}
\partial(h_1 + \phi_*(c_1) + \theta_L(u_1)) &= \partial(h_1) + \phi_*(\partial(c_1)) + \theta_L(\partial(u_1)) \\
&= \theta_L(h_0) + \phi_*(\theta_K(c_0)) + \theta_L(h_0 + \phi_*(c_0)) \\
&= \theta_L(h_0) + \theta_L(\phi_*(c_0)) + \theta_L(h_0) + \theta_L(\phi_*(c_0)) = \mathbf{0}.
\end{aligned}
$$

Here we have used $\theta_L \circ \phi_* = \phi_* \circ \theta_K$, which holds by the assumption that ϕ is equivariant. It follows that there is a 2-chain u_2 with $\partial(u_2) = h_1 + \phi_*(c_1) + \theta_L(u_1)$. The analogous computation shows $h_2 + \phi_*(c_2) + \theta_L(u_2)$ is a cycle and so there is a 3-chain with $\partial(u_3) = h_2 + \phi_*(c_2) + \theta_L(u_2)$. Continuing in this manner, we find j-chains u_j with $\partial(u_j) = h_{j-1} + \phi_*(c_{j-1}) + \theta_L(u_{j-1})$ for $1 \le j \le n$ ($u_0 = \mathbf{0}$). By construction, $h_n + \phi_*(c_n) + \theta_L(u_n)$ is an n-cycle in $C_n(L; \mathbb{F}_2)$ and so it is homologous to either $\theta_L(h_n)$ or to $\mathbf{0}$ since $H_n(L; \mathbb{F}_2) \cong \mathbb{F}_2[\{[\theta_L(h_n)]\}]$. In either case, $\theta_L(h_n + \phi_*(c_n) + \theta_L(u_n)) = \theta_L(h_n) + \phi_*(\theta_K(c_n))$ is homologous to $\mathbf{0}$. Let $c = \theta_K(c_n)$; then

$$
\partial(c) = \partial(\theta_K(c_n)) = \theta_K(\partial(c_n)) = \theta_K(\theta_K(c_{n-1})) = \mathbf{0},
$$

and so $[c] \in H_n(K; \mathbb{F}_2)$ satisfies $H(\phi)([c]) = [\phi_*(c)] = [\phi_*(\theta_K(c_n))] = [\theta_L(h_n)]$ and $[\bar{\nu}_*(c)] = [\bar{\nu}_*(\theta_K(c_n))] = [\theta_K(c_n)] = [c]$, so $H(\nu)([c]) = [c]$. □

Corollary 11.14. *There are no equivariant mappings* $F \colon (S^n, a) \to (S^m, a)$ *when* $n > m$.

Proof. The homology of S^n has no nonzero classes in $H_j(S^n; \mathbb{F}_2)$ for $1 \le j \le m$, and so, if there were an equivariant mapping $F \colon S^n \to S^m$, the conclusion of Theorem 11.13 would fail. □

The Borsuk-Ulam Theorem is the case $m = n - 1$. There are many proofs of the Borsuk-Ulam Theorem, as well as remarkable applications in diverse parts of mathematics. The interested reader should consult [**56**] for more details (and a great read).

Exercises

1. Suppose X and Y are triangulable spaces that are homotopy equivalent. Show that $H_p(X; \mathbb{F}_2) \cong H_p(Y; \mathbb{F}_2)$ for all p. The

notion of contiguous simplicial mappings (Theorem 10.21) plays a big role here.

2. Use the homotopy invariance of homology to compute the homology of the Möbius band.

3. The projective plane $\mathbb{R}\mathrm{P}^2$ is modeled by an explicit simplicial complex, pictured in Chapter 10. The combinatorial data allow one to construct the sequence of boundary homomorphisms

$$C_2(\mathbb{R}\mathrm{P}^2; \mathbb{F}_2) \xrightarrow{\partial} C_1(\mathbb{R}\mathrm{P}^2; \mathbb{F}_2) \xrightarrow{\partial} C_0(\mathbb{R}\mathrm{P}^2; \mathbb{F}_2) \to \{\mathbf{0}\}.$$

This may be boiled down to a pair of matrices whose ranks determine the homology. Use this formulation to compute $H_j(\mathbb{R}\mathrm{P}^2; \mathbb{F}_2)$ for all j.

4. If L is a subcomplex of a simplicial complex K, $L \subset K$, then we can define the homology of the pair (K, L) by setting

$$C_p(K, L; \mathbb{F}_2) = C_p(K; \mathbb{F}_2)/C_p(L; \mathbb{F}_2).$$

Show that the boundary operator on the chains on K and L defines a boundary operator on the quotient vector space $C_p(K, L; \mathbb{F}_2)$. Then $H_p(K, L; \mathbb{F}_2)$ is the quotient of the kernel of the boundary operator by the image of the boundary operator. Compute $H_p(K, L; \mathbb{F}_2)$ for all p when K is a cylinder $S^1 \times [0, 1]$ and L is its boundary (a pair of circles), and when K is the Möbius band, and L its boundary.

5. A path through a simplex can be deformed to pass only through the subcomplex of edges (1-simplices) of the simplex. Because a simplex is convex, this gives a homotopy between the path and its deformation. Use this idea to define a mapping $\pi_1(|K|, v_0) \to H_1(K; \mathbb{F}_2)$ that sends a loop based at a vertex v_0 to a 1-chain in K. Show that the mapping so defined is a group homomorphism. What happens in the case that $|K| \cong S^1$?

Where from here?

The diligent reader who has mastered the better part of this book is ready for a great deal more. I have restricted my attention to particular spaces and particular methods in order to focus on the question of the topological invariance of dimension. The quick route to the proof of invariance of dimension left a lot of the landscape unexplored. In particular, the question of dimension can be posed more generally, for which a rich theory has been developed. The interested reader can consult [38] for the classic treatment, and the articles of Johnson [40] and Dauben [17] for a history of its development. For topics in the general history of topology, there is the collection of essays edited by James [39], the three volume handbook edited by Aull and Lowen [80], and the sweeping account of Dieudonné [19].

Where to go next is best answered by recommending some texts for which the reader is now ready.

A far broader treatment of the topics in this book can be found in the books of Munkres, [61] and [62]. Enthusiasts of point-set topology (Chapters 1–6) will find a rich vein there. Other treatments of point-set topics can be found in [42] and [34], and there is the collection of sometimes surprising counterexamples to sharpen point-set topological intuition found in [75].

The fundamental group is thoroughly presented in the classic book of Massey [55] and in the lectures of Lima [51]. A deeper exploration of the idea of covering spaces leads to a topological setting for a Galois correspondence, which has been a fruitful analogy.

For the purposes of ease of exposition toward our main goal, I introduced homology with coefficients in $\mathbb{F}_2$. It is possible to define homology with other coefficients, $H_*(X; A)$ for A an abelian group, and for arbitrary topological spaces, singular homology, by developing the properties of simplices with more care. This is the usual place to start a graduate course in algebraic topology. I recommend [55], [62], [30], [32], [73], and [16] for these topics. With more subtle chains, many interesting geometric results can be proved.

The most important examples of topological spaces throughout the history of topology are manifolds. These are spaces which are locally homeomorphic to open sets in $\mathbb{R}^n$ for which the methods of the Calculus play a principal role. The interface between topology and analysis is subtle and made clear on manifolds. This is the subject of differential topology, treated in [60], [21], and [54].

I did not treat some of the other classical topological topics in this book about which the reader may curious. On the subject of knots, the books of Colin Adams [1] and Livingston [52] are good introductions. The problem of classifying all surfaces is presented in [55] and [7]. Geometric topics, like the Poincaré index theorem, are a part of classical topology, and can be read about in [34].

Finally, the notation $\pi_0(X)$ and $\pi_1(X)$ hints at a sequence of groups, $\pi_n(X)$, known as the higher homotopy groups of a space X. The iterative definition, introduced by Hurewicz [37], is

$$\pi_n(X) = \pi_{n-1}(\Omega(X, x_0)).$$

For example, the second homotopy group of X is the fundamental group of the based loop space on X. The properties of these groups and their computation for particular spaces X is a difficult problem. Some aspects of this problem are developed in [15], [57], [58], and [73].

To the budding topologist, I wish many exciting discoveries.

Bibliography

1. Adams, Colin, The Knot Book, American Mathematical Society, Providence, RI, 2004.

2. Agoston, M. K., Algebraic Topology, a first course, Marcel Dekker, Inc., New York, NY, 1976.

3. Alexander, J. W., A proof of Jordan's theorem about a simple closed curve, Annals of Math. **21**(1920), 180–184.

4. Alexandroff, P. S., Stetige Abbildinger kompakter Räume, Math. Annalen **96**(1926), 555–573.

5. Alexandroff, P. S., Hopf, H., Topologie I, Springer-Verlag, Berlin, 1935.

6. Alexandroff, P. S., Combinatorial Topology, Dover Publications, Mineola, NY, 1998.

7. Armstrong, M.A., Basic Topology, UTM Series, Springer-Verlag, Berlin-Heidlelberg-New York, 1997.

8. Björner, A., Topological methods, in Handbook of Combinatorics, Vol. 1, 2, Edited by R. L. Graham, M. Grötschel and L. Lovász, Elsevier, Amsterdam, 1995, 1819–1872.

9. Borsuk, K., Drei Sätze über die n-dimensionale euklidische Sphäre, Fund. Math. **20**(1933), 177–190.

10. Brouwer, L. E. J., Beweis der Invarianz der Dimensionzahl, Math. Ann. **70**(1910), 161–165.

11. Brouwer, L. E. J., Über Abbildungen von Mannigfaltigkeiten, Math. Ann. **71**(1911), 97–115.

12. Burde, G., Zieschang, H., Knots, Walter de Gruyter, Berlin-New York, 1985.

13. Cantor, G., Ein Beitrag zur Mannigfaltigkeitslehre, J. für reine und ang. Math. **84**(1878), 242–258.

14. Cox, R. H., A proof of the Schroeder-Berstein Theorem, American Mathematical Monthly **75**(1968), 508.

15. Croom, F. H., Basic Concepts of Algebraic Topology (Undergraduate Texts in Mathematics), Springer-Verlag, Berlin-Heidelberg-New York, 1978.

16. Crossley, M. D., Essential Topology Series: Springer Undergraduate Mathematics Series, Springer-Verlag, Berlin-Heidelberg-New York, 2005.

17. Dauben, J. W., The invariance of dimension: Problems in the early development of set theory and topology, Historia Math. **2**(1975), 273–288.

18. Day, Mahlon M., Normed linear spaces, Third edition, Ergebnisse der Mathematik und ihrer Grenzgebiete, Band 21, Springer-Verlag, New York-Heidelberg, 1973.

19. Dieudonné, J., A History of Algebraic and Differential Topology, 1900–1960, Birkäuser, Boston-Basel, 1989.

20. Dold, A., Lectures on Algebraic Topology, Springer-Verlag, Heidelberg, 1972.

21. Dubrovin, B.A., Fomenko, A.T., Novikov, S.P., Modern Geometry–Methods and Applications, Part I. The Geometry of Surfaces, Transformation Groups, and Fields; Part II. The Geometry and Topology of Manifolds; Part III. Introduction to Homology Theory; Springer-Verlag, New York, Translated by Robert G. Burns, 1984, 1985, 1990.

22. Eilenberg, S, Mac Lane, S., General theory of natural equivalences, Trans. Amer. Math. Soc. **58**(1945), 231–294.

23. Eilenberg, S., Steenrod, N., Foundations of Algebraic Topology, Princeton University Press, Princeton, NJ, 1952.

24. Eilenberg, S., Zilber, J., On products of complexes, Amer. J. Math. **75**(1953), 200–204.

25. Epple, M., Orbits of asteroids, a braid, and the first link invariant, Math. Intelligencer **20**(1998), 45–52.

26. Fine, B., Rosenberger, G., The Fundamental Theorem of Algebra, Undergraduate Texts in Mathematics series, Springer-Verlag, New York, 1997.

27. Fréchet, M., Sur quelques points du calcul functionnel, Rend. Circ. Mat. Palermo **22**(1906), 1–74.

28. Gamelin, T. W., Greene, R. E., Introduction to Topology, Second edition, Dover Publ., Inc., Mineola, NY, 1999.

29. Giblin, Peter, Graphs, Surfaces and Homology, Chapman and Hall, London, 1977.

30. Greenberg, M. J., Harper, J. R., Algebraic Topology, a first course, Benjamin-Cummings, Reading, MA, 1981.

31. Hardy, G. H., Wright, E. M., An Introduction to the Theory of Numbers, fifth edition, Oxford University Press, Oxford, England, 1980.

32. Hatcher, Allen, Algebraic Topology, Cambridge University Press, New York, NY, 2001.

33. Hausdorff, F., Grundzüge der Mengentheorie, Chelsea Publ. Co., New York, NY, 1949.

34. Henle, M., A Combinatorial Introduction to Topology, Dover Publ. Inc., Mineola, NY, 1994.

35. Hilton, P. J., Wylie, S., Homology theory: An introduction to algebraic topology, Cambridge University Press, New York, NY, 1960.

36. Horn, R. A., Johnson, C. R., Topics in Matrix Analysis, Cambridge University Press, New York, 1991.

37. Hurewicz, W., Beiträge zur Topologie der Deformationen, I: Höherdimensionalen Homotopiegruppen; II: Homotopie- und Homologiegruppen; III: Klassen und Homologietypen von Abbildungen; IV: Asphärische Räume, Proc. Akad. Wetensch. Amsterdam **38**(1935), 112-119, 521-528, **39**(1936), 117-126, 215-224.

38. Hurewicz, W., Wallman, H., Dimension Theory, Princeton University Press, Princeton, NY, 1941.

39. James, I. M., editor, History of Topology, Elsevier, Amsterdam, 1999.

40. Johnson, Dale M., Prelude to dimension theory: The geometrical investigations of Bernard Bolzano, Arch. History Exact Sci. **17**(1977), 262-295; The problem of the invariance of dimension in the growth of modern topology. I, Arch. Hist. Exact Sci. **20**(1979), 97-188; The problem of the invariance of dimension in the growth of modern topology. II, Arch. Hist. Exact Sci. **25**(1981), 85-267.

41. Jordan, Camille, Cours d'Analyse de l'Ecole polytechnique, Gauthier-Villars, Paris, 1882.

42. Kahn, D. W., Topology: An Introduction to the Point-Set and Algebraic Areas, Dover Publ. Inc., Mineola, NY, 1995.

43. Kaku, Michio, Hyperspace: A Scientific Odyssey Through Parallel Universes, Time Warps and the Tenth Dimension, Oxford University Press, 1994.

44. Kelley, J. L., The Tychonoff product theorem implies the axiom of choice, Fund. Math. **37**(1950), 75-76.

45. Klein, F., Vergleichende Betrachtungen über neuere geometrische Forshungen, Math. Ann. **43**(1893), 63–100.

46. Lakatos, I., Proofs and Refutations: The Logic of Mathematical Discovery, Cambridge University Press, Cambridge, UK, 1976.

47. Lebesgue, H., Sur la non-applicabilité de deux domaines appartenant respectivement à des espaces à n et $n + p$ dimensions, Math. Ann. **71**(1910), 166–168.

48. Lefschetz, S., Topology, Amer. Math. Soc. Colloq. Publ. **12**, Providence, RI, 1930.

49. Lefschetz, S., Algebraic Topology, Amer. Math. Soc. Colloq. Publ. **27**, Providence, RI, 1942.

50. Lefschetz, S., The early development of algebraic topology, Bol. Soc. Bras. Matem. **1**(1970), 1–48.

51. Lima, Elon Lages, Fundamental Groups and Covering Spaces, translated by Jonas Gomes, A.K. Peters, Natick, MA, 2003.

52. Livingston, C., Knot Theory, Carus Monograph Series, Mathematical Association of America, Washington, DC, 1996.

53. Mac Lane, S., Categories for the Working Mathematician, second edition, GTM Series 5, Springer-Verlag, Berlin-Heidelberg-New York, 1998.

54. Madsen, I., Tornehave, J., From Calculus to Cohomology, Cambridge University Press, Cambridge, 1997.

55. Massey, W. S., A Basic Course in Algebraic Topology, GTM 127, Springer-Verlag, Berlin-Heidelberg-New York, 1991.

56. Matoušek, J., Using the Borsuk-Ulam Theorem, Springer-Verlag, Berlin-Heidelberg, 2003.

57. Maunder, C. R. F., Algebraic Topology, Dover Publ., Mineola, NY, 1996.

58. May, J. P., A Concise Course in Algebraic Topology, Chicago Lectures in Mathematics, Univ. Chicago Press, Chicago, IL, 1999.

59. McCleary, J., Geometry from a Differentiable Viewpoint, Cambridge University Press, New York, NY, 1997.

60. Milnor, J. W., Topology from a Differentiable Viewpoint, Princeton University Press, Princeton, NJ, 1997.

61. Munkres, James R., Topology: A first course, second edition, Prentice-Hall, Inc., Englewood Cliffs, NJ, 1999.

62. Munkres, James, Elements of Algebraic Topology, Addison-Wesley Publ. Co., Menlo Park, CA, 1984.

63. Newman, M. H. A., Elements of the Topology of Plane Sets of Points, Cambridge University Press, Cambridge, UK, 1939.

64. Ostrowski, A., Über den ersten und vierten Gausschen Beweis des Fundamentalsatzes der Algebra, in Gauss Werke, Vol. 10, Part 2, Abh. 3.

65. Peano, G., Sur une courbe, qui rempli une aire plane, Math. Ann. **36**(1890), 157–160.

66. Poincaré, H., Analysis Situs, J. Ecole Polytechnique **1**(1895), 1–121.

67. Poincaré, H., Science and Hypothesis, Walter Scott Publishing, London, 1905.

68. Pont, Jean-Claude, La topologie algébrique des origines à Poincaré, Préface de René Taton, Bibliothèque de Philosophie Contemporaine, Presses Universitaires de France, Paris, 1974.

69. Rolfsen, D., Knots and Links, AMS Chelsea Publ., Providence, RI, 2003.

70. Royden, H. L., Real Analysis, Third edition, Macmillan Publ. Co., New York, 1988.

71. Sagan, H., Space-Filling Curves, Springer-Verlag, New York, 1994.

72. Schmidt, E., Über den Jordanschen Kurvensatz, Sitzber. Akad. Berlin, (1923), 318–329.

73. Spanier, E. H., Algebraic Topology, Springer-Verlag, Berlin-Heidelberg-New York, 1994.

74. Stanton, D., White, D., Constructive Combinatorics, Undergraduate Texts in Mathematics, Springer-Verlag, New York, 1986.

75. Steen, L. A., Seebach, J. A., Counterexamples in Topology, Dover Publ. Inc., Mineola, NY, 1995.

76. Uspensky, J. V., Theory of Equations, McGraw-Hill Book Co., New York, NY, 1948.

77. Vassiliev, V. A., Introduction to Topology (Student Mathematical Library, V. 14), A. Sossinski (Translator), American Mathematical Society, Providence, RI, 2001.

78. Walker, J. W., A homology version of the Borsuk-Ulam theorem, Amer. Math. Monthly **90**(1983), 466–468.

79. Wall, C. T. C., A Geometric Introduction to Topology, Addison-Wesley Publ. Co., Reading, MA, 1972.

80. Aull, C. E., Lowen, R. (eds.), Handbook of the History of General Topology, Kluwer Academic Publ., Dordrecht, vol. 1 (1997), vol. 2 (1998), vol. 3 (2001).

Notation Index

Subject Index

Titles in This Series

For a complete list of titles in this series, visit the AMS Bookstore at **www.ams.org/bookstore/**.